ISBN-13: 978-1-951619-07-7

This is the Solution Guide to the book "Real Analysis for Beginners."

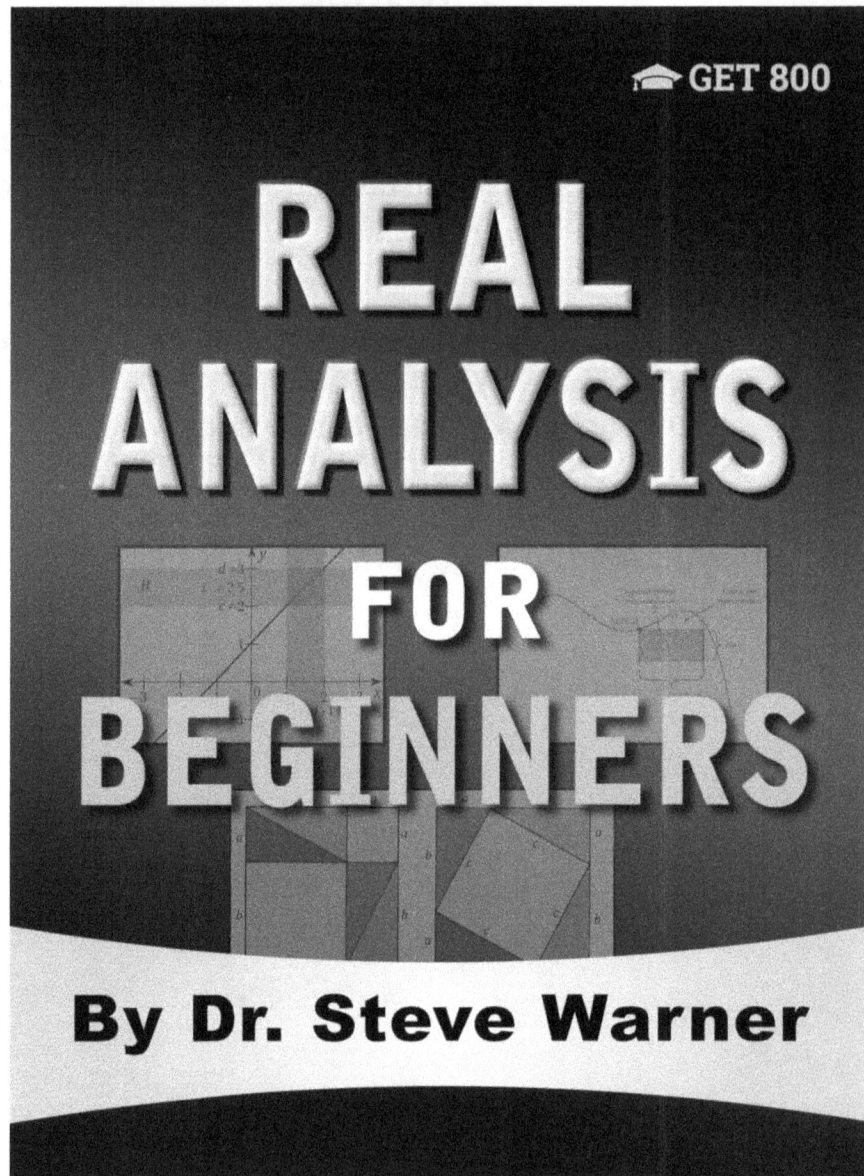

Also Available from Dr. Steve Warner

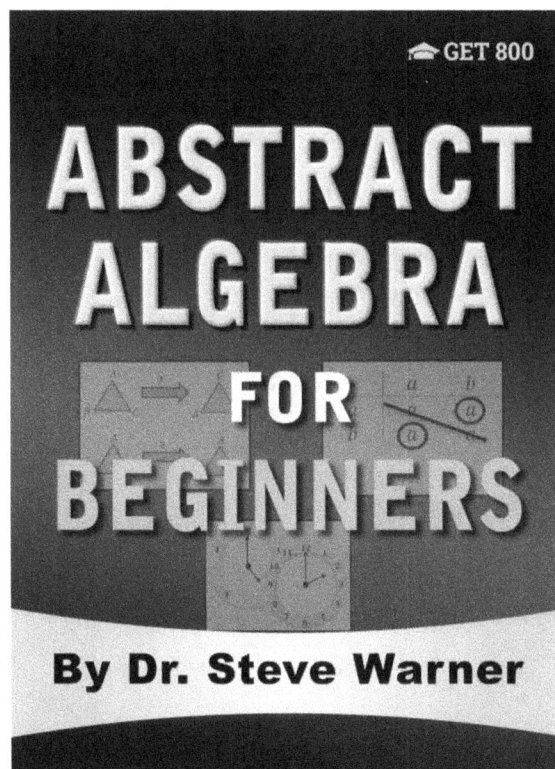

CONNECT WITH DR. STEVE WARNER

www.facebook.com/SATPrepGet800

www.youtube.com/TheSATMathPrep

www.twitter.com/SATPrepGet800

www.linkedin.com/in/DrSteveWarner

www.pinterest.com/SATPrepGet800

Also Available from Dr. Steve Warner

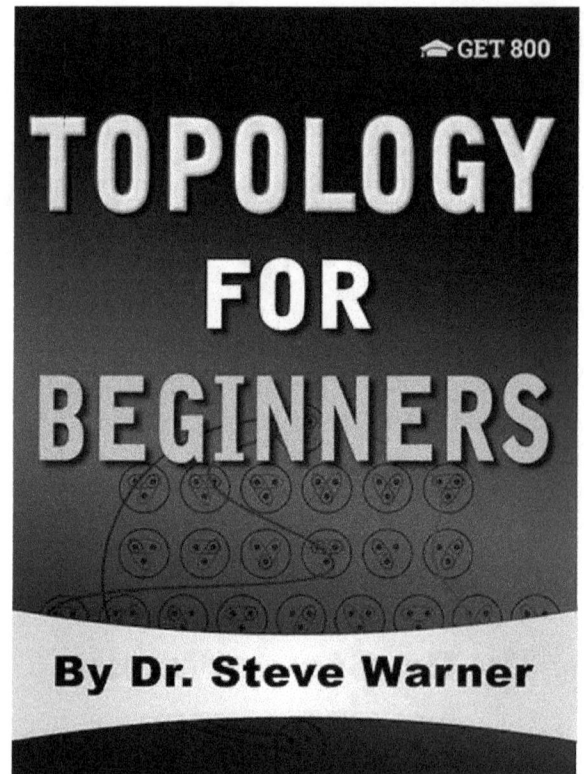

CONNECT WITH DR. STEVE WARNER

www.facebook.com/SATPrepGet800

www.youtube.com/TheSATMathPrep

www.twitter.com/SATPrepGet800

www.linkedin.com/in/DrSteveWarner

www.pinterest.com/SATPrepGet800

Real Analysis for Beginners

Solution Guide

Dr. Steve Warner

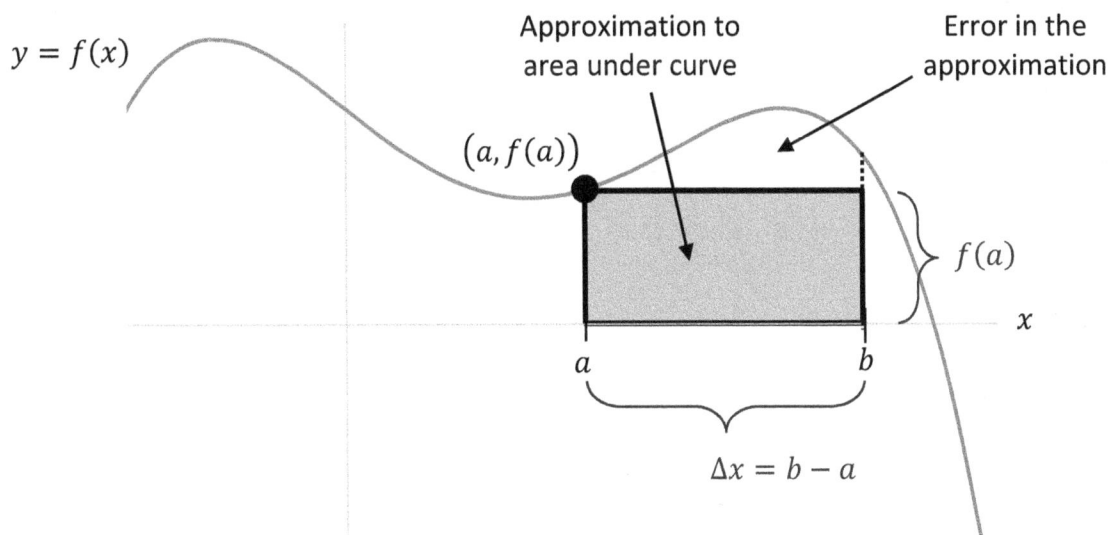

$y = f(x)$

Approximation to area under curve

Error in the approximation

$(a, f(a))$

$f(a)$

a

b

x

$\Delta x = b - a$

Table of Contents

Problem Set 1   7

Problem Set 2   21

Problem Set 3   26

Problem Set 4   36

Problem Set 5   49

Problem Set 6   55

Problem Set 7   72

Problem Set 8   80

Problem Set 9   90

Problem Set 10   95

Problem Set 11   103

Problem Set 12   121

Problem Set 13   128

Problem Set 14   141

Problem Set 15   148

Problem Set 16   158

About the Author   *168*

Books by Dr. Steve Warner   169

Problem Set 1

1. Determine whether each of the following statements is true or false:

   (i) $9 \in \{9\}$

   (ii) $c \in \{a, b, c\}$

   (iii) $-3 \in \{3\}$

   (iv) $\frac{2}{3} \in \mathbb{Z}$

   (v) $-18 \in \mathbb{N}$

   (vi) $\frac{3}{19} \in \mathbb{Q}$

   (vii) $\emptyset \subseteq \{0, 1, 2\}$

   (viii) $\{\delta\} \subseteq \{\delta, \Delta\}$

   (ix) $\{f, g, h\} \subseteq \{f, g, h\}$

   (x) $\{a, b, \{c, d\}\} \subseteq \{a, b, c, d\}$

Solutions:

   (i) $\{9\}$ has exactly 1 element, namely 9. So, $9 \in \{9\}$ is **true**.

   (ii) $\{a, b, c\}$ has exactly 3 elements, namely a, b, and c. In particular, $c \in \{a, b, c\}$ is **true**.

   (iii) $\{3\}$ has exactly 1 element, namely 3. So, $-3 \notin \{3\}$. Therefore, $-3 \in \{3\}$ is **false**.

   (iv) $\mathbb{Z} = \{\ldots, -4, -3, -2, -1, 0, 1, 2, 3, 4, \ldots\}$. So, $\frac{2}{3} \in \mathbb{Z}$ is **false**.

   (v) $\mathbb{N} = \{0, 1, 2, 3, \ldots\}$. Therefore, $-18 \in \mathbb{N}$ is **false**.

   (vi) Since $3, 19 \in \mathbb{Z}$ and $19 \neq 0$, $\frac{3}{19} \in \mathbb{Q}$ is **true**.

   (vii) The empty set is a subset of every set. So, $\emptyset \subseteq \{0, 1, 2\}$ is **true**.

   (viii) The only element of $\{\delta\}$ is δ. Since δ is also an element of $\{\delta, \Delta\}$, $\{\delta\} \subseteq \{\delta, \Delta\}$ is **true**.

   (ix) Every set is a subset of itself. So, $\{f, g, h\} \subseteq \{f, g, h\}$ is **true**.

   (x) $\{c, d\} \in \{a, b, \{c, d\}\}$, but $\{c, d\} \notin \{a, b, c, d\}$. So, $\{a, b, \{c, d\}\} \subseteq \{a, b, c, d\}$ is **false**.

2. Determine the cardinality of each of the following sets:

   (i) $\{0, 1, 2\}$

   (ii) $\{a, b, c, d, e, f\}$

   (iii) $\{1, 2, \ldots, 99\}$

   (iv) $\left\{\frac{1}{2}, \frac{1}{3}, \ldots, \frac{1}{11}\right\}$

Solutions:

   (i) $|\{0, 1, 2\}| = \textbf{3}$.

   (ii) $|\{a, b, c, d, e, f\}| = \textbf{6}$.

   (iii) $|\{1, 2, ..., 99\}| = \textbf{99}$.

   (iv) $\left|\left\{\frac{1}{2}, \frac{1}{3}, ..., \frac{1}{11}\right\}\right| = \textbf{10}$.

3. List the elements of $\{k, x, t\} \times \{5, 6\}$.

Solution: $(k, 5), (k, 6), (x, 5), (x, 6), (t, 5), (t, 6)$

4. Let $A = \{0\}$. Evaluate (i) A^2; (ii) A^3; (iii) $\mathcal{P}(A)$.

Solutions:

   (i) $\{0\}^2 = \{0\} \times \{0\} = \{(0, 0)\}$.

   (ii) $\{0\}^3 = \{0\} \times \{0\} \times \{0\} = \{(0, 0, 0)\}$.

   (iii) $\mathcal{P}(\{0\}) = \{\emptyset, \{0\}\}$

5. Let $A = \{a, b, \Delta, \delta\}$ and $B = \{b, c, \delta, \gamma\}$. Determine each of the following:

   (i) $A \cup B$

   (ii) $A \cap B$

   (iii) $A \setminus B$

   (iv) $B \setminus A$

   (v) $A \, \Delta \, B$

Solutions:

   (i) $A \cup B = \{a, b, c, \Delta, \delta, \gamma\}$.

   (ii) $A \cap B = \{b, \delta\}$.

   (iii) $A \setminus B = \{a, \Delta\}$

   (iv) $B \setminus A = \{c, \gamma\}$

   (v) $A \, \Delta \, B = \{a, \Delta\} \cup \{c, \gamma\} = \{a, c, \Delta, \gamma\}$

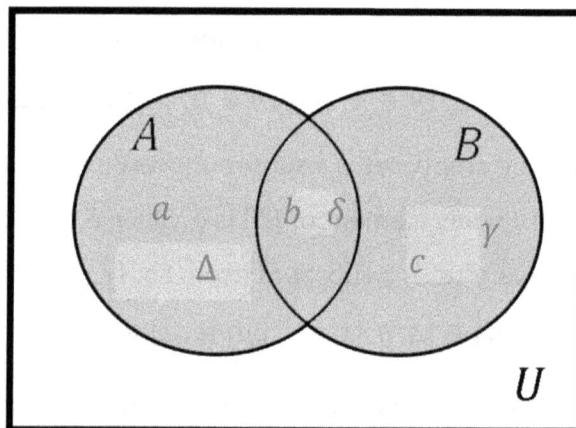

8

6. Draw Venn diagrams for $(A \setminus B) \setminus C$ and $A \setminus (B \setminus C)$. Are these two sets equal for all sets A, B, and C? If so, prove it. If not, provide a counterexample.

Solution:

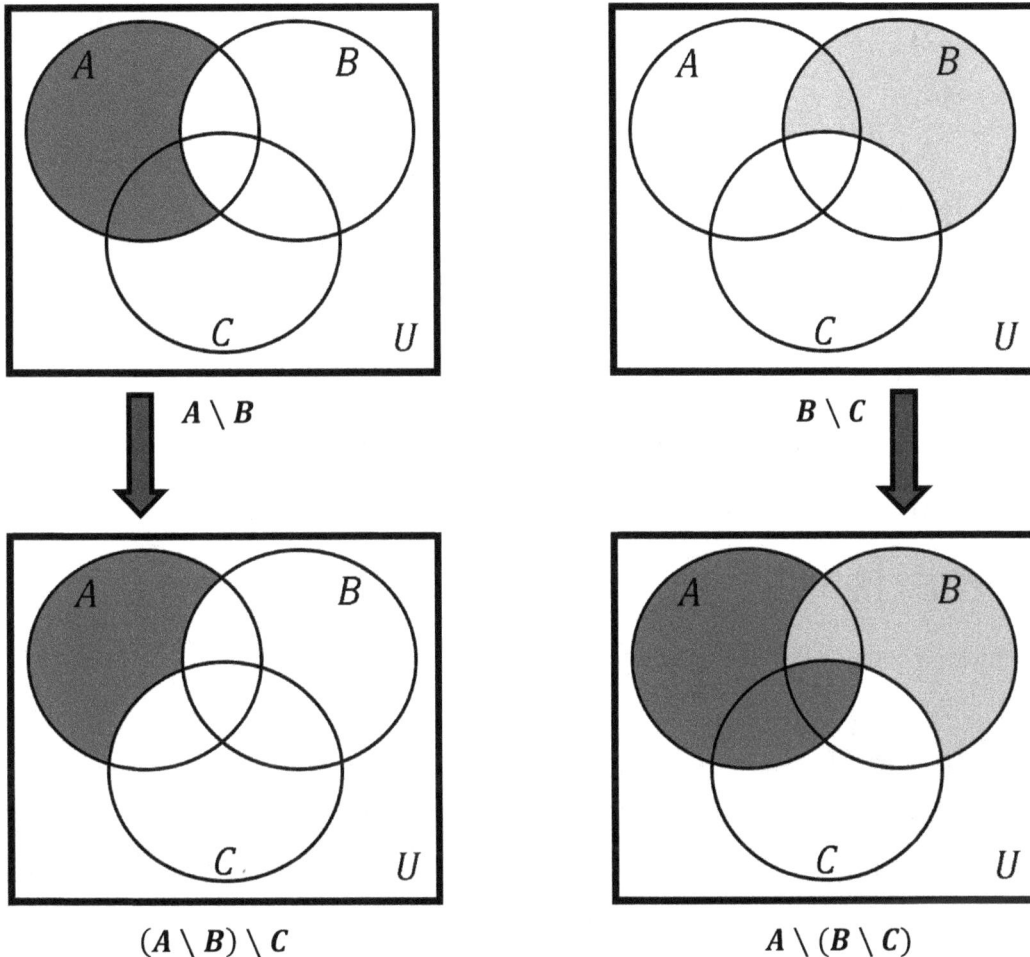

$A \setminus B$

$B \setminus C$

$(A \setminus B) \setminus C$

$A \setminus (B \setminus C)$

From the Venn diagrams, it looks like $(A \setminus B) \setminus C \subseteq A \setminus (B \setminus C)$, but $(A \setminus B) \setminus C \neq A \setminus (B \setminus C)$.

Let's come up with a counterexample. Let $A = \{1, 2\}$, $B = \{1, 3\}$, and $C = \{1, 4\}$. Then we have $(A \setminus B) \setminus C = \{2\} \setminus \{1, 4\} = \{2\}$ and $A \setminus (B \setminus C) = \{1, 2\} \setminus \{3\} = \{1, 2\}$.

We see that $(A \setminus B) \setminus C \neq A \setminus (B \setminus C)$.

Note: Although it was not asked in the question, let's prove that $(A \setminus B) \setminus C \subseteq A \setminus (B \setminus C)$. Let $x \in (A \setminus B) \setminus C$. Then $x \in A \setminus B$ and $x \notin C$. Since $x \in A \setminus B$, $x \in A$ and $x \notin B$. In particular, $x \in A$. Since $x \notin B$, $x \notin B \setminus C$ (because if $x \in B \setminus C$, then $x \in B$). So, we have $x \in A$ and $x \notin B \setminus C$. Therefore, $x \in A \setminus (B \setminus C)$. Since $x \in (A \setminus B) \setminus C$ was arbitrary, $(A \setminus B) \setminus C \subseteq A \setminus (B \setminus C)$. $\quad\square$

7. Compute the power set of each of the following sets:

   (i) $\emptyset$

   (ii) $\{2\}$

   (iii) $\{a, b\}$

   (iv) $\{\emptyset, \{\emptyset\}\}$

   (v) $\{\{\emptyset\}\}$

Solutions:

   (i) $\mathcal{P}(\emptyset) = \{\emptyset\}$

   (ii) $\mathcal{P}(\{2\}) = \{\emptyset, \{2\}\}$

   (iii) $\mathcal{P}(\{a, b\}) = \{\emptyset, \{a\}, \{b\}, \{a, b\}\}$

   (iv) $\mathcal{P}(\{\emptyset, \{\emptyset\}\}) = \{\emptyset, \{\emptyset\}, \{\{\emptyset\}\}, \{\emptyset, \{\emptyset\}\}\}$

   (v) $\mathcal{P}(\{\{\emptyset\}\}) = \{\emptyset, \{\{\emptyset\}\}\}$

8. Determine whether each of the following statements is true or false:

   (i) $2 \in \emptyset$

   (ii) $\emptyset \in \{a, b\}$

   (iii) $\emptyset \in \emptyset$

   (iv) $\emptyset \in \{\emptyset, \{\emptyset\}\}$

   (v) $\{\emptyset\} \in \emptyset$

   (vi) $\{\emptyset\} \in \{\emptyset\}$

   (vii) $\emptyset \subseteq \emptyset$

   (viii) $\emptyset \subseteq \{\emptyset\}$

   (ix) $\{\emptyset\} \subseteq \emptyset$

   (x) $\{\emptyset\} \subseteq \{\emptyset\}$

   (xi) $\mathbb{Q} \subseteq \mathbb{C}$

   (xii) $3 \in \{2k \mid k = 1, 2, 3, 4, 5, 6\}$

Solutions:

   (i)   The empty set has no elements. So, $x \in \emptyset$ is false for any x. In particular, $2 \in \emptyset$ is **false**.

   (ii) $\{a, b\}$ has exactly 2 elements, namely a and b. In particular, $\emptyset \notin \{a, b\}$. So, $\emptyset \in \{a, b\}$ is **false**.

   (iii) The empty set has no elements. So, $x \in \emptyset$ is false for any x. In particular, $\emptyset \in \emptyset$ is **false**.

(iv)   The set $\{\emptyset, \{\emptyset\}\}$ has exactly 2 elements, namely $\emptyset$ and $\{\emptyset\}$. In particular, $\emptyset \in \{\emptyset, \{\emptyset\}\}$ is **true**.

(v)   The empty set has no elements. So, $x \in \emptyset$ is false for any x. In particular, $\{\emptyset\} \in \emptyset$ is **false**.

(vi)   The set $\{\emptyset\}$ has 1 element, namely $\emptyset$. Since $\{\emptyset\} \neq \emptyset$, $\{\emptyset\} \in \{\emptyset\}$ is **false**.

(vii)   The empty set is a subset of every set. So, $\emptyset \subseteq X$ is true for any X. In particular, $\emptyset \subseteq \emptyset$ is **true**. (This can also be done by using the fact that every set is a subset of itself.)

(viii) Again, (as in (vii)), $\emptyset \subseteq X$ is true for any X. In particular, $\emptyset \subseteq \{\emptyset\}$ is **true**.

(ix)   The only subset of $\emptyset$ is $\emptyset$. So, $\{\emptyset\} \subseteq \emptyset$ is **false**.

(x)   Every set is a subset of itself. So, $\{\emptyset\} \subseteq \{\emptyset\}$ is **true**.

(xi)   Since $\mathbb{Q} \subseteq \mathbb{R}$ and $\mathbb{R} \subseteq \mathbb{C}$, by Theorem 1.14, $\mathbb{Q} \subseteq \mathbb{C}$ is **true**.

(xii) $\{2k \mid k = 1, 2, 3, 4, 5, 6\} = \{2, 4, 6, 8, 10, 12\}$. So, $3 \notin \{2k \mid k = 1, 2, 3, 4, 5, 6\}$. Therefore, $3 \in \{2k \mid k = 1, 2, 3, 4, 5, 6\}$ is **false**.

9. Determine the cardinality of each of the following sets:

   (i) $\{a, a, b, c, d, d, d\}$

   (ii) $\{\{1, 2\}, \{3,4,5\}\}$

   (iii) $\{5, 6, 7, \ldots, 2122, 2123\}$

Solutions:

(i) $\{a, a, b, c, d, d, d\} = \{a, b, c, d\}$. Therefore, $|\{a, a, b, c, d, d, d\}| = |\{a, b, c, d\}| = \mathbf{4}$.

(ii) $\{\{1, 2\}, \{3,4,5\}\}$ consists of the 2 elements $\{1, 2\}$ and$\{3,4,5\}$. So, $\left|\{\{1, 2\}, \{3,4,5\}\}\right| = \mathbf{2}$.

(iii) $|\{5, 6, 7, \ldots, 2122, 2123\}| = 2123 - 5 + 1 = \mathbf{2119}$.

Note: For number (iii), we used the fence-post formula (see Notes 3 and 4 after Example 1.7).

10. Compute $\{a, b\}^4$.

Solution: $\{a, b\}^4 = \{a, b\} \times \{a, b\} \times \{a, b\} \times \{a, b\} = \{(a, a, a, a), (a, a, a, b), (a, a, b, a), (a, a, b, b),$
$(a, b, a, a), (a, b, a, b), (a, b, b, a), (a, b, b, b), (b, a, a, a), (b, a, a, b), (b, a, b, a), (b, a, b, b),$
$(b, b, a, a), (b, b, a, b), (b, b, b, a), (b, b, b, b)\}$,

11. Let $A = \{\emptyset, \{\emptyset, \{\emptyset\}\}\}$ and $B = \{\emptyset, \{\emptyset\}\}$. Compute each of the following:

   (i) $A \cup B$

   (ii) $A \cap B$

   (iii) $A \setminus B$

   (iv) $B \setminus A$

   (v) $A \Delta B$

11

Solutions:

(i) $A \cup B = \{\emptyset, \{\emptyset\}, \{\emptyset, \{\emptyset\}\}\}$.

(ii) $A \cap B = \{\emptyset\}$.

(iii) $A \setminus B = \{\{\emptyset, \{\emptyset\}\}\}$

(iv) $B \setminus A = \{\{\emptyset\}\}$

(v) $A \triangle B = \{\{\emptyset, \{\emptyset\}\}\} \cup \{\{\emptyset\}\} = \{\{\emptyset\}, \{\emptyset, \{\emptyset\}\}\}$

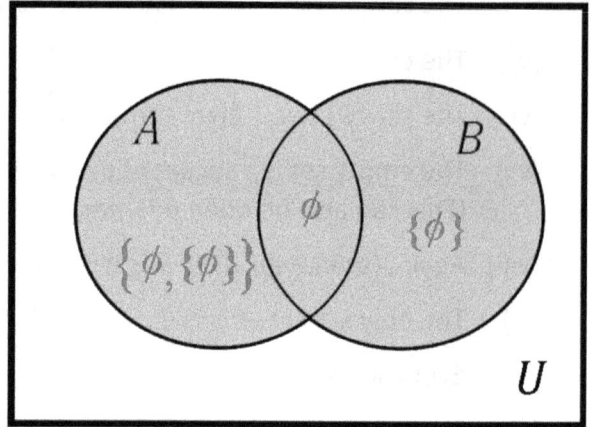

12. Prove the following:

    (i)   The operation of forming unions is commutative.

    (ii)   The operation of forming intersections is commutative.

    (iii)   The operation of forming intersections is associative.

Proofs:

(i) Let A and B be sets. Then $x \in A \cup B$ if and only if $x \in A$ or $x \in B$ if and only if $x \in B$ or $x \in A$ if and only if $x \in B \cup A$. Since x was arbitrary, we have shown $\forall x (x \in A \cup B \leftrightarrow x \in B \cup A)$. Therefore, $A \cup B = B \cup A$. So, the operation of forming unions is commutative.   □

(ii) Let A and B be sets. Then $x \in A \cap B$ if and only if $x \in A$ and $x \in B$ if and only if $x \in B$ and $x \in A$ if and only if $x \in B \cap A$. Since x was arbitrary, we have $\forall x (x \in A \cap B \leftrightarrow x \in B \cap A)$. Therefore, $A \cap B = B \cap A$. So, the operation of forming intersections is commutative.   □

(iii) Let A, B, and C be sets. Then $x \in (A \cap B) \cap C$ if and only if $x \in A \cap B$ and $x \in C$ if and only if $x \in A$, $x \in B$ and $x \in C$ if and only if $x \in A$ and $x \in B \cap C$ if and only if $x \in A \cap (B \cap C)$. Since x was arbitrary, we have shown $\forall x (x \in (A \cap B) \cap C \leftrightarrow x \in A \cap (B \cap C))$.

Therefore, we have shown that $(A \cap B) \cap C = A \cap (B \cap C)$. So, the operation of forming intersections is associative.   □

LEVEL 3

13. Determine the cardinality of each of the following sets:

    (i) $\{\{\{a, b\}\}\}$

    (ii) $\{\{0, 1\}, 0, \{0\}, \{0, \{0, 1, 2\}\}\}$

    (iii) $\{a, \{a\}, \{a, a\}, \{a, a, a, a\}, \{a, a, \{a\}\}, \{a, \{a\}, \{a\}\}\}$

Solutions:

(i) The only element of $\{\{\{a, b\}\}\}$ is $\{\{a, b\}\}$. So, $\left|\{\{\{a, b\}\}\}\right| = 1$.

(ii)   The elements of $\big\{\{0,1\},0,\{0\},\{0,\{0,1,2\}\}\big\}$ are $\{0,1\}$, 0, $\{0\}$, and $\{0,\{0,1,2\}\}$. So, we see that $\Big|\big\{\{0,1\},0,\{0\},\{0,\{0,1,2\}\}\big\}\Big| = \mathbf{4}.$

(iii)   We have: $\big\{a,\{a\},\{a,a\},\{a,a,a,a\},\{a,a,\{a\}\},\{a,\{a\},\{a\}\}\big\}$

$$= \big\{a,\{a\},\{a\},\{a\},\{a,\{a\}\},\{a,\{a\}\}\big\}$$

$$= \big\{a,\{a\},\{a,\{a\}\}\big\}.$$

So, $\Big|\big\{a,\{a\},\{a,a\},\{a,a,a,a\},\{a,a,\{a\}\},\{a,\{a\},\{a\}\}\big\}\Big| = \Big|\big\{a,\{a\},\{a,\{a\}\}\big\}\Big| = \mathbf{3}.$

14. How many subsets does $\{a,b,c,d\}$ have? Draw a tree diagram for the subsets of $\{a,b,c,d\}$.

Solution: $|\{a,b,c,d\}| = 4$. Therefore, $\{a,b,c,d\}$ has $2^4 = \mathbf{16}$ subsets. We can also say that the size of the power set of $\{a,b,c,d\}$ is 16, that is, $|\mathcal{P}(\{a,b,c,d\})| = 16$. Here is a tree diagram.

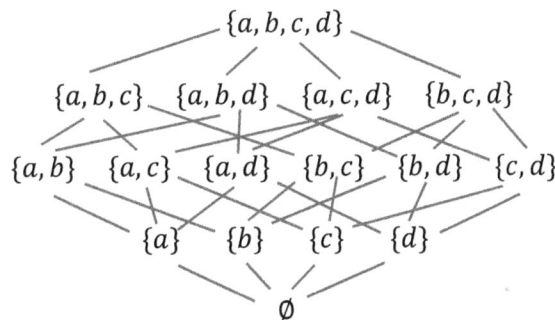

15. Let A, B, C, D, and E be sets such that $A \subseteq B$, $B \subseteq C$, $C \subseteq D$, and $D \subseteq E$. Prove that $A \subseteq E$.

Proof: Suppose that A, B, C, D, and E are sets such that $A \subseteq B$, $B \subseteq C$, $C \subseteq D$, and $D \subseteq E$. Since $A \subseteq B$ and $B \subseteq C$, by Theorem 1.14, we have $A \subseteq C$. Since $A \subseteq C$ and $C \subseteq D$, again by Theorem 1.14, we have $A \subseteq D$. Finally, since $A \subseteq D$ and $D \subseteq E$, once again by Theorem 1.14, we have $A \subseteq E$.   □

16. Let A, B, C, and D be sets with $A \subseteq B$ and $C \subseteq D$. Prove that $A \times C \subseteq B \times D$.

Proof: Let A, B, C, and D be sets with $A \subseteq B$ and $C \subseteq D$ and let $(x,y) \in A \times C$. Then $x \in A$ and $y \in C$. Since $x \in A$ and $A \subseteq B$, $x \in B$. Since $y \in C$ and $C \subseteq D$, $y \in D$. Therefore, $(x,y) \in B \times D$. Since $(x,y) \in A \times C$ was arbitrary, $A \times C \subseteq B \times D$.   □

17. Prove or provide a counterexample:

   (i)   Every pairwise disjoint set of sets is disjoint.

   (ii)   Every disjoint set of sets is pairwise disjoint.

Solutions:

(i)   This is **false**. Let $A = \{1\}$ and let $X = \{A\}$. X is pairwise disjoint, but $\cap X = A = \{1\} \neq \emptyset$.

   However, the following slightly modified statement is **true**: "Every pairwise disjoint set of sets consisting of at least two sets is disjoint."

13

Let X be a pairwise disjoint set of sets with at least two sets, say $A, B \in X$. Suppose towards contradiction that $x \in \cap X$. Then $x \in A$ and $x \in B$. So, $x \in A \cap B$. But $A \cap B = \emptyset$ because X is pairwise disjoint. This contradiction shows that the statement $x \in \cap X$ is false. Therefore, X is disjoint. □

(ii) This is **false**. Let $A = \{0, 1\}$, $B = \{1, 2\}$, $C = \{0, 2\}$, and $X = \{A, B, C\}$. Then X is disjoint because $\cap X = A \cap B \cap C = \{0, 1\} \cap \{1, 2\} \cap \{0, 2\} = \{1\} \cap \{0, 2\} = \emptyset$. However, X is **not** pairwise disjoint because $A \cap B = \{0, 1\} \cap \{1, 2\} = \{1\} \neq \emptyset$.

18. Let A and B be sets. Prove that $A \cap B \subseteq A$.

Proof: Suppose that A and B are sets and let $x \in A \cap B$. Then $x \in A$ and $x \in B$. In particular, $x \in A$. Since x was an arbitrary element of A, we have shown that every element of $A \cap B$ is an element of A. That is, $\forall x (x \in A \cap B \rightarrow x \in A)$ is true. Therefore, $A \cap B \subseteq A$. □

LEVEL 4

19. Determine whether each of the following statements is true or false:

(i) $c \in \{a, \{c\}\}$

(ii) $\{\Delta\} \in \{\delta, \Delta\}$

(iii) $\{1\} \in \{1, a, 2, b\}$

(iv) $\emptyset \in \{\{\emptyset\}\}$

(v) $\{\{\emptyset\}\} \in \emptyset$

Solutions:

(i) The set $\{a, \{c\}\}$ has exactly 2 elements, namely a and $\{c\}$. So, $c \in \{a, \{c\}\}$ is **false**.

(ii) The set $\{\delta, \Delta\}$ has exactly 2 elements, namely δ and Δ. So, $\{\Delta\} \in \{\delta, \Delta\}$ is **false**.

(iii) The set $\{1, a, 2, b\}$ has exactly 4 elements, namely $1, a, 2$, and b. So, $\{1\} \in \{1, a, 2, b\}$ is **false**.

(iv) The set $\{\{\emptyset\}\}$ has exactly 1 element, namely $\{\emptyset\}$. Since $\emptyset$ is not equal to $\{\emptyset\}$, $\emptyset \in \{\{\emptyset\}\}$ is **false**.

(v) The empty set has no elements. So, $x \in \emptyset$ is false for any x. In particular, $\{\{\emptyset\}\} \in \emptyset$ is **false**.

20. Let A, B, C, and D be sets. Determine if each of the following statements is true or false. If true, provide a proof. If false, provide a counterexample.

(i) $(A \times B) \cap (C \times D) = (A \cap C) \times (B \cap D)$

(ii) $(A \times B) \cup (C \times D) = (A \cup C) \times (B \cup D)$

Solutions:

(i) This is **true**.

Proof: $(x, y) \in (A \times B) \cap (C \times D)$ if and only if $(x, y) \in A \times B$ and $(x, y) \in C \times D$ if and only if $x \in A$, $y \in B$, $x \in C$, and $y \in D$ if and only if $x \in A \cap C$ and $y \in B \cap D$ if and only if $(x, y) \in (A \cap C) \times (B \cap D)$. Therefore, $(A \times B) \cap (C \times D) = (A \cap C) \times (B \cap D)$. □

(ii)   This is **false**. If $A = \{0\}, B = \{1\}, C = \{2\}, D = \{3\}$, then $A \times B = \{(0, 1)\}, C \times D = \{(2, 3)\}$, and so, $(A \times B) \cup (C \times D) = \{(0, 1), (2, 3)\}$. Also, $A \cup C = \{0, 2\}$, $B \cup D = \{1, 3\}$, and so, $(A \cup C) \times (B \cup D) = \{(0, 1), (0, 3), (2, 1), (2, 3)\}$. Since $(2, 1) \in (A \cup C) \times (B \cup D)$, but $(2, 1) \notin (A \times B) \cup (C \times D)$, we see that $(A \times B) \cup (C \times D) \neq (A \cup C) \times (B \cup D)$.

21. Prove that $B \subseteq A$ if and only if $A \cap B = B$.

Proof: Suppose that $B \subseteq A$. By part (ii) of Problem 12 and Problem 18, $A \cap B = B \cap A \subseteq B$. Let $x \in B$. Since $B \subseteq A$, $x \in A$. Therefore, $x \in A$ and $x \in B$. So, $x \in A \cap B$. Since x was an arbitrary element of B, we have shown that every element of B is an element of $A \cap B$. That is, $\forall x(x \in B \rightarrow x \in A \cap B)$. Therefore, $B \subseteq A \cap B$. Since $A \cap B \subseteq B$ and $B \subseteq A \cap B$, it follows that $A \cap B = B$.

Now, suppose that $A \cap B = B$ and let $x \in B$. Then $x \in A \cap B$. So, $x \in A$ and $x \in B$. In particular, $x \in A$. Since x was an arbitrary element of B, we have shown that every element of B is an element of A. That is, $\forall x(x \in B \rightarrow x \in A)$. Therefore, $B \subseteq A$. □

22. Let A, B, and C be sets. Prove each of the following:

   (i) $A \cap (B \cup C) = (A \cap B) \cup (A \cap C)$.

   (ii) $A \cup (B \cap C) = (A \cup B) \cap (A \cup C)$.

   (iii) $C \setminus (A \cup B) = (C \setminus A) \cap (C \setminus B)$.

   (iv) $C \setminus (A \cap B) = (C \setminus A) \cup (C \setminus B)$.

Proofs:

   (i) $x \in A \cap (B \cup C) \Leftrightarrow x \in A$ and $x \in B \cup C \Leftrightarrow x \in A$ and either $x \in B$ or $x \in C \Leftrightarrow x \in A$ and $x \in B$ or $x \in A$ and $x \in C \Leftrightarrow x \in A \cap B$ or $x \in A \cap C \Leftrightarrow x \in (A \cap B) \cup (A \cap C)$. □

   (ii) $x \in A \cup (B \cap C) \Leftrightarrow x \in A$ or $x \in B \cap C \Leftrightarrow$ either $x \in A$ or we have both $x \in B$ and $x \in C \Leftrightarrow$ we have both $x \in A$ or $x \in B$ and $x \in A$ or $x \in C \Leftrightarrow x \in A \cup B$ and $x \in A \cup C \Leftrightarrow x \in (A \cup B) \cap (A \cup C)$. □

   (iii) $x \in C \setminus (A \cup B) \Leftrightarrow x \in C$ and $x \notin A \cup B \Leftrightarrow x \in C$ and $x \notin A$ and $x \notin B \Leftrightarrow x \in C$ and $x \notin A$ and $x \in C$ and $x \notin B \Leftrightarrow x \in C \setminus A$ and $x \in C \setminus B \Leftrightarrow x \in (C \setminus A) \cap (C \setminus B)$. □

   (iv) $x \in C \setminus (A \cap B) \Leftrightarrow x \in C$ and $x \notin A \cap B \Leftrightarrow x \in C$ and $x \notin A$ or $x \notin B \Leftrightarrow x \in C$ and $x \notin A$ or $x \in C$ and $x \notin B \Leftrightarrow x \in C \setminus A$ or $x \in C \setminus B \Leftrightarrow x \in (C \setminus A) \cup (C \setminus B)$. □

Notes: Let's let p, q, and r be the statements $x \in A$, $x \in B$, and $x \in C$, respectively.

(1) In (i) above, the statement "$x \in A$ and either $x \in B$ or $x \in C$" can be written $p \wedge (q \vee r)$. It can easily be shown that this is equivalent to $(p \wedge q) \vee (p \wedge r)$. In words, this is the statement "$x \in A$ and $x \in B$ or $x \in A$ and $x \in C$." Here it needs to be understood that the word "and" takes precedence over the word "or."

Similarly, we can use the logical equivalence $p \lor (q \land r) \equiv (p \lor q) \land (p \lor r)$ to help understand the proof of (ii).

(2) The equivalences $p \land (q \lor r) \equiv (p \land q) \lor (p \land r)$ and $p \lor (q \land r) \equiv (p \lor q) \land (p \lor r)$ are known as the **distributive laws**.

The rules $A \cap (B \cup C) = (A \cap B) \cup (A \cap C)$ and $A \cup (B \cap C) = (A \cup B) \cap (A \cup C)$ are also known as the **distributive laws**.

(3) To clarify (iii) and (iv), note that $\neg(p \lor q) \equiv \neg p \land \neg q$ and $\neg(p \land q) \equiv \neg p \lor \neg q$ (these equivalences can be easily checked). These two equivalences are known as **De Morgan's laws**. For (iii), we can use the logical equivalence $\neg(p \lor q) \equiv \neg p \land \neg q$ with p the statement $x \in A$ and q the statement $x \in B$ to get

$$x \notin A \cup B \equiv \neg x \in A \cup B \equiv \neg(x \in A \lor x \in B) \equiv \neg(p \lor q) \equiv \neg p \land \neg q \text{ (by De Morgan's law)}$$
$$\equiv \neg x \in A \land \neg x \in B \equiv x \notin A \land x \notin B.$$

So, the statement "$x \in C$ and $x \notin A \cup B$" is equivalent to $x \in C \land x \notin A \land x \notin B$.

Similarly, we can use the logical equivalence $\neg(p \land q) \equiv \neg p \lor \neg q$ to see that the statement "$x \in C$ and $x \notin A \cap B$" is equivalent to "$x \in C$ and $x \notin A$ or $x \notin B$."

(4) The rules $C \setminus (A \cup B) = (C \setminus A) \cap (C \setminus B)$ and $C \setminus (A \cap B) = (C \setminus A) \cup (C \setminus B)$ are also known as **De Morgan's laws**.

LEVEL 5

23. Let A and B be sets with $B \subseteq A$. Determine if the following are true or false. If true, provide a proof. If false, provide a counterexample.

    (i) $B \in \mathcal{P}(A)$

    (ii) $B \subseteq \mathcal{P}(A)$

    (iii) $\mathcal{P}(B) \in \mathcal{P}(A)$

    (iv) $\mathcal{P}(B) \subseteq \mathcal{P}(A)$

Solutions:

  (i) **True**.

     Proof: Let A and B be sets with $B \subseteq A$. Then $B \in \mathcal{P}(A)$ (by the definition of $\mathcal{P}(A)$).   □

  (ii) **False**.

     Counterexample: Let $A = \{a, b\}$ and $B = \{a\}$. Then $\mathcal{P}(A) = \{\emptyset, \{a\}, \{b\}, \{a, b\}\}$. Now, $a \in B$, but $a \notin \mathcal{P}(A)$. Therefore, $B \nsubseteq \mathcal{P}(A)$.

(iii) **False**.

Counterexample: Let $A = \{0, 1\}$ and $B = \{1\}$. Then $\mathcal{P}(A) = \{\emptyset, \{0\}, \{1\}, \{0, 1\}\}$ and $\mathcal{P}(B) = \{\emptyset, \{1\}\} \notin \mathcal{P}(A)$.

(iv) **True**.

Proof: Let A and B be sets with $B \subseteq A$ and let $X \in \mathcal{P}(B)$. Then $X \subseteq B$ Since $X \subseteq B$ and $B \subseteq A$, by the transitivity of $\subseteq$, $X \subseteq A$. So, $X \in \mathcal{P}(A)$. Since $X \in \mathcal{P}(B)$ was arbitrary, $\mathcal{P}(B) \subseteq \mathcal{P}(A)$. $\square$

24. Let $P(x)$ be the property $x \notin x$. Prove that $\{x | P(x)\}$ cannot be a set.

Solution: Suppose toward contradiction that $A = \{x \mid x \notin x\}$ is a set. Then $A \in A$ if and only if $A \notin A$. So, $p \leftrightarrow \neg p$ is true, where p is the statement $A \in A$. However, $p \leftrightarrow \neg p$ is always false. This is a contradiction. So, A is not a set. $\square$

Notes: (1) This is our first **proof by contradiction**. A proof by contradiction works as follows:

1. We assume the negation of what we are trying to prove.

2. We use a logically sound argument to derive a statement which is false.

3. Since the argument is logically sound, the only possible error is our original assumption. Therefore, the negation of our original assumption must be true.

In this problem we are trying to prove that $A = \{x \mid x \notin x\}$ **is not** a set. The negation of this statement is that $A = \{x \mid x \notin x\}$ **is** a set. We then use only the definition of A to get the false statement $A \in A \leftrightarrow \neg A \in A$. Since the argument was logically valid, our initial assumption must have been incorrect, and therefore, A is not a set.

(2) The contradiction that occurs here is known as **Russell's Paradox**. This contradiction shows that we need to be careful about how we define a set. A naïve definition would be that a set is any object that has the form $\{x | P(x)\}$, where $P(x)$ is an arbitrary property (by property, we mean a **first-order property**—this is a property defined using the connectives $\land$, $\lor$, $\rightarrow$, and $\leftrightarrow$, the quantifiers $\forall$ and $\exists$, and the relations $=$ and $\in$). As we see in this problem, that "definition" of a set leads to a contradiction. Instead, we call $\{x | P(x)\}$ a **class**. Every set is a class, but not every class is a set. A class that is not a set is called a **proper class**. For example, $\{x | x \notin x\}$ is a proper class.

25. Let $A = \{a, b, c, d\}$, $B = \{X \mid X \subseteq A \land d \notin X\}$, and $C = \{X \mid X \subseteq A \land d \in X\}$. Show that there is a natural one-to-one correspondence between the elements of B and the elements of C. Then generalize this result to a set with $n + 1$ elements for $n > 0$.

Solution: We define the one-to-one correspondence as follows: If $Y \in B$, then Y is a subset of A that does not contain d. Let Y_d be the set that contains the same elements as Y, but with d thrown in. Then the correspondence $Y \rightarrow Y_d$ is a one-to-one correspondence. We can see this correspondence in the table below.

Elements of B	Elements of C
$\emptyset$	$\{d\}$
$\{a\}$	$\{a,d\}$
$\{b\}$	$\{b,d\}$
$\{c\}$	$\{c,d\}$
$\{a,b\}$	$\{a,b,d\}$
$\{a,c\}$	$\{a,c,d\}$
$\{b,c\}$	$\{b,c,d\}$
$\{a,b,c\}$	$\{a,b,c,d\}$

For the general result, we start with a set A with $n+1$ elements, and we let d be some element from A. Define B and C the same way as before: $B = \{X \mid X \subseteq A \land d \notin X\}$, and $C = \{X \mid X \subseteq A \land d \in X\}$. Also, as before, if $Y \in B$, then Y is a subset of A that does not contain d. Let Y_d be the set that contains the same elements as Y, but with d thrown in. Then the correspondence $Y \to Y_d$ is a one-to-one correspondence.

Notes: (1) B consists of the subsets of A that do not contain the element d, while C consists of the subsets of A that do contain d.

(2) Observe that in the case where $A = \{a,b,c,d\}$, B and C each have $8 = 2^3$ elements. Also, there is no overlap between B and C (they have no elements in common). So, we have a total of $8 + 8 = 16$ elements. Since there are exactly $2^4 = 16$ subsets of A, we see that we have listed every subset of A.

(3) We could also do the computation in Note 2 as follows: $2^3 + 2^3 = 2 \cdot 2^3 = 2^1 \cdot 2^3 = 2^{1+3} = 2^4$. It's nice to see the computation this way because it mimics the computation we will do in the more general case. In case your algebra skills are not that strong, here is an explanation of each step:

Adding the same thing to itself is equivalent to multiplying that thing by 2. For example, 1 apple plus 1 apple is 2 apples. Similarly, $1x + 1x = 2x$. This could be written more briefly as $x + x = 2x$. Replacing x by 2^3 gives us $2^3 + 2^3 = 2 \cdot 2^3$ (the first equality in the computation above).

Next, by definition, $x^1 = x$. So, $2^1 = 2$. Therefore, we can rewrite $2 \cdot 2^3$ as $2^1 \cdot 2^3$.

Now, 2^3 means to multiply 2 by itself 3 times. So, $2^3 = 2 \cdot 2 \cdot 2$. Thus, $2^1 \cdot 2^3 = 2 \cdot 2 \cdot 2 \cdot 2 = 2^4$. This leads to the rule of exponents which says that if you multiply two expressions with the same base, you can add the exponents. So, $2^1 \cdot 2^3 = 2^{1+3} = 2^4$.

(4) In the more general case, B and C each have 2^n elements. The reason for this is that A has $n+1$ elements. When we remove the element d from A, the resulting set has n elements, and therefore, 2^n subsets. B consists of precisely the subsets of this new set (A with d removed), and so, B has exactly 2^n elements. The one-to-one correspondence $Y \to Y_d$ shows that C has the same number of elements as B. Therefore, C also has 2^n elements.

(5) In the general case, there is still no overlap between B and C. It follows that the total number of elements when we combine B and C is $2^n + 2^n = 2 \cdot 2^n = 2^1 \cdot 2^n = 2^{1+n} = 2^{n+1}$. See Note 3 above for an explanation as to how all this algebra works.

(6) By a **one-to-one correspondence** between the elements of B and the elements of C, we mean a pairing where we match each element of B with exactly one element of C so that each element of C is matched with exactly one element of B. The table given in the solution above provides a nice example of such a pairing.

(7) In the case where $A = \{a, b, c, d\}$, B consists of all the subsets of $\{a, b, c\}$. In other words, $B = \{X \mid X \subseteq \{a, b, c\}\} = \mathcal{P}(\{a, b, c\})$.

A description of C is a bit more complicated. It consists of the subsets of $\{a, b, c\}$ with d thrown into them. We could write this as $C = \{X \cup \{d\} \mid X \subseteq \{a, b, c\}\}$.

(5) In the general case, we can write $K = A \setminus \{d\}$ (this is the set consisting of all the elements of A, except d). We then have $B = \{X \mid X \subseteq K\} = \mathcal{P}(K)$ and $C = \{X \cup \{d\} \mid X \subseteq K\} = \mathcal{P}(A) \setminus \mathcal{P}(K)$.

26. Let X be a nonempty set of sets. Prove the following:

    (i)   For all $A \in X$, $A \subseteq \cup X$.

    (ii)   For all $A \in X$, $\cap X \subseteq A$.

Proofs:

    (i)   Let X be a nonempty set of sets, let $A \in X$, and let $x \in A$. Then there is $B \in X$ such that $x \in B$ (namely A). So, $x \in \cup X$. Since x was an arbitrary element of A, we have shown that $A \subseteq \cup X$. Since A was an arbitrary element of X, we have shown that for all $A \in X$, we have $A \subseteq \cup X$.   □

    (ii)   Let X be a nonempty set of sets, let $A \in X$, and let $x \in \cap X$. Then for every $B \in X$, we have $x \in B$. In particular, $x \in A$ (because $A \in X$). Since x was an arbitrary element of $\cap X$, we have shown that $\cap X \subseteq A$. Since A was an arbitrary element of X, we have shown that for all $A \in X$, we have $\cap X \subseteq A$.   □

27. Let A be a set and let X be a nonempty set of sets. Prove each of the following:

    (i) $A \cap \cup X = \cup \{A \cap B \mid B \in X\}$

    (ii) $A \cup \cap X = \cap \{A \cup B \mid B \in X\}$

    (iii) $A \setminus \cup X = \cap \{A \setminus B \mid B \in X\}$

    (iv) $A \setminus \cap X = \cup \{A \setminus B \mid B \in X\}$.

Proofs:

    (i) $x \in A \cap \cup X \Leftrightarrow x \in A$ and $x \in \cup X \Leftrightarrow x \in A$ and there is a $B \in X$ with $x \in B \Leftrightarrow x \in A \cap B$ for some $B \in X \Leftrightarrow x \in \cup \{A \cap B \mid B \in X\}$.   □

(ii) $x \in A \cup \cap X \Leftrightarrow x \in A$ or $x \in \cap X \Leftrightarrow x \in A$ or $x \in B$ for every $B \in X \Leftrightarrow x \in A \cup B$ for every $B \in X \Leftrightarrow x \in \cap \{A \cup B \mid B \in X\}$. □

(iii) $x \in A \setminus \cup X \Leftrightarrow x \in A$ and $x \notin \cup X \Leftrightarrow x \in A$ and $x \notin B$ for every $B \in X \Leftrightarrow x \in A \setminus B$ for every $B \in X \Leftrightarrow x \in \cap \{A \setminus B \mid B \in X\}$. □

(iv) $x \in A \setminus \cap X \Leftrightarrow x \in A$ and $x \notin \cap X \Leftrightarrow x \in A$ and $x \notin B$ for some $B \in X \Leftrightarrow x \in A \setminus B$ for some $B \in X \Leftrightarrow x \in \cup \{A \setminus B \mid B \in X\}$. □

Note: The rules in (i) and (ii) are known as the **generalized distributive laws** and the rules in (iii) and (iv) are known as the **generalized De Morgan's laws.**

Problem Set 2

LEVEL 1

1. Determine if each of the following sets is an interval of real numbers:

   (i) $A = \{x \in \mathbb{R} \mid 3 \leq x \leq 7\}$

   (ii) $B = \{x \in \mathbb{Q} \mid x < -205\}$

   (iii) $C = \mathbb{R}^+$

   (iv) $D = \{x \in \mathbb{R} \mid x \geq -16\}$

   (v) $E = \mathbb{R} \setminus \{0\}$

Solutions:

(i)   **Yes**: $A = [3, 7]$

(ii) **No**

(iii) **Yes**: $C = (0, \infty)$

(iv) **Yes**: $D = [-16, \infty)$

(v)   **No**: for example, $-1 < 0 < 1$, $-1, 1 \in \mathbb{R} \setminus \{0\}$, but $0 \notin \mathbb{R} \setminus \{0\}$.

2. Let $C = (-\infty, 2]$ and $D = (-1, 3]$. Compute each of the following:

   (i) $C \cup D$

   (ii) $C \cap D$

   (iii) $C \setminus D$

   (iv) $D \setminus C$

   (v) $C \Delta D$

Solutions:

(vi) $C \cup D = (-\infty, 3]$

(vii) $C \cap D = (-1, 2]$

(viii) $C \setminus D = (-\infty, -1]$

(ix) $D \setminus C = (2, 3]$

(x) $C \Delta D = (-\infty, -1] \cup (2, 3]$

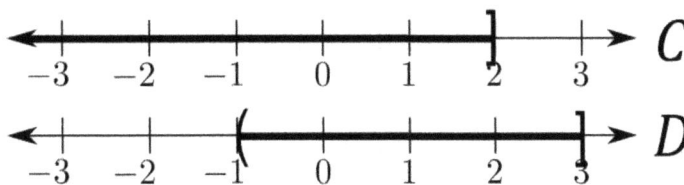

3. Find all partitions of the three-element set $\{a, b, c\}$ and the four-element set $\{a, b, c, d\}$.

Solution: The partitions of $\{a, b, c\}$ are $\{\{a\}, \{b\}, \{c\}\}$, $\{\{a\}, \{b, c\}\}$, $\{\{b\}, \{a, c\}\}$, $\{\{c\}, \{a, b\}\}$, and $\{\{a, b, c\}\}$.

21

The partitions of $\{a, b, c, d\}$ are $\{\{a\}, \{b\}, \{c\}, \{d\}\}$, $\{\{a\}, \{b\}, \{c, d\}\}$, $\{\{a\}, \{c\}, \{b, d\}\}$, $\{\{a\}, \{d\}, \{b, c\}\}$, $\{\{b\}, \{c\}, \{a, d\}\}$, $\{\{b\}, \{d\}, \{a, c\}\}$, $\{\{c\}, \{d\}, \{a, b\}\}$, $\{\{a, b\}, \{c, d\}\}$, $\{\{a, c\}, \{b, d\}\}$, $\{\{a, d\}, \{b, c\}\}$, $\{\{a, b, c\}, \{d\}\}$, $\{\{a, b, d\}, \{c\}\}$, $\{\{a, c, d\}, \{b\}\}$, $\{\{b, c, d\}, \{a\}\}$, and $\{\{a, b, c, d\}\}$.

4. Let $A = \{1, 2, 3, 4\}$ and let $R = \{(1, 1), (1, 3), (2, 2), (2, 4), (3, 1), (3, 3), (4, 2), (4, 4)\}$. Note that R is an equivalence relation on A. Find the equivalence classes of R.

Solution: The equivalence classes of R are $\{1, 3\}$ and $\{2, 4\}$.

LEVEL 2

5. Find the domain, range, and field of each of the following relations:
    (i) $R = \{(a, b), (c, d), (e, f), (f, a)\}$
    (ii) $S = \{(2k, 2t + 1) \mid k, t \in \mathbb{Z}\}$

Solutions:

(i) $\operatorname{dom} R = \{a, c, e, f\}$; $\operatorname{ran} R = \{a, b, d, f\}$; field $R = \{a, b, c, d, e, f\}$

(ii) $\operatorname{dom} S = \{2k \mid k \in \mathbb{Z}\} = 2\mathbb{Z} = \mathbb{E}$; $\operatorname{ran} S = \{2t + 1 \mid t \in \mathbb{Z}\} = \mathbb{O}$; field $S = \mathbb{Z}$

6. Prove that for each $n \in \mathbb{Z}^+$, $\equiv_n$ (see part 4 of Example 2.12) is an equivalence relation on $\mathbb{Z}$.

Proof: Let $a \in \mathbb{Z}$. Then $a - a = 0 = n \cdot 0$. So, $n | a - a$. Therefore, $a \equiv_n a$, and so, $\equiv_n$ is reflexive. Let $a, b \in \mathbb{Z}$ and suppose that $a \equiv_n b$. Then $n | b - a$. So, there is $k \in \mathbb{Z}$ such that $b - a = nk$. Thus, $a - b = -(b - a) = -nk = n(-k)$. Since $k \in \mathbb{Z}$, $-k \in \mathbb{Z}$. So, $n | a - b$, and therefore, $b \equiv_n a$. So, $\equiv_n$ is symmetric. Let $a, b, c \in \mathbb{Z}$ with $a \equiv_n b$ and $b \equiv_n c$. Then $n | b - a$ and $n | c - b$. So, there are $j, k \in \mathbb{Z}$ such that $b - a = nj$ and $c - b = nk$. So, $c - a = (c - b) + (b - a) = nk + nj = n(k + j)$. Since the sum of two integers is an integer, $k + j \in \mathbb{Z}$. Therefore, $n | c - a$. So, $a \equiv_n c$. Thus, $\equiv_n$ is transitive. Since $\equiv_n$ is reflexive, symmetric, and transitive, $\equiv_n$ is an equivalence relation on $\mathbb{Z}$. $\square$

LEVEL 3

7. Prove that there do not exist sets A and B such that the relation $<$ on $\mathbb{R}$ is equal to $A \times B$.

Proof: Suppose toward contradiction that $< = A \times B$ for some sets A and B. Since $1 < 2$ and $2 < 3$, we have $(1, 2), (2, 3) \in A \times B$. Since $(2, 3) \in A \times B$, $2 \in A$. Since $(1, 2) \in A \times B$, $2 \in B$. Therefore, $(2, 2) \in A \times B$. Since $< = A \times B$, $2 < 2$, contradicting that $<$ is antireflexive. Therefore, there do not exist sets A and B such that $<$ is equal to $A \times B$. $\square$

8. Let X be a set of equivalence relations on a nonempty set A. Prove that $\bigcap X$ is an equivalence relation on A.

Proof: Let X be a set of equivalence relations on a nonempty set A. Let $x \in A$ and let $R \in X$. Since R is reflexive, $(x, x) \in R$. Since $R \in X$ was arbitrary, $\forall R \in X \big((x, x) \in R\big)$. So, $(x, x) \in \bigcap X$. Since $x \in A$ was arbitrary, $\bigcap X$ is reflexive.

Let $(x, y) \in \cap X$ and let $R \in X$. Then $(x, y) \in R$. Since R is an equivalence relation, R is symmetric. Therefore, $(y, x) \in R$. Since $R \in X$ was arbitrary, $\forall R \in X((y, x) \in R)$. So, $(y, x) \in \cap X$. Since $(x, y) \in \cap X$ was arbitrary, $\cap X$ is symmetric.

Let $(x, y), (y, z) \in \cap X$ and let $R \in X$. Then $(x, y), (y, z) \in R$. Since R is an equivalence relation, R is transitive. Therefore, $(x, z) \in R$. Since $R \in X$ was arbitrary, $\forall R \in X((x, z) \in R)$. So, $(x, z) \in \cap X$. Since $(x, y), (y, z) \in \cap X$ was arbitrary, $\cap X$ is transitive.

Since $\cap X$ is reflexive, symmetric, and transitive, $\cap X$ is an equivalence relation. $\quad\square$

LEVEL 4

9. Let $R = \{(x, y) \in \mathbb{R} \times \mathbb{R} \mid x - y \in \mathbb{Z}\}$. Prove that R is an equivalence relation on $\mathbb{R}$ and describe the equivalence classes of R.

Proof: If $x \in \mathbb{R}$, then $x - x = 0 \in \mathbb{Z}$. So, xRx, and therefore, R is reflexive. If xRy, then $x - y \in \mathbb{Z}$. It follows that $y - x = -(x - y) \in \mathbb{Z}$. So, yRx, and therefore, R is symmetric. If xRy and yRz, then $x - y \in \mathbb{Z}$ and $y - z \in \mathbb{Z}$. It follows that $x - z = (x - y) + (y - z) \in \mathbb{Z}$ (because the sum of two integers is an integer). So, xRz, and therefore, R is transitive. Since R is reflexive, symmetric, and transitive, R is an equivalence relation.

For each $r \in \mathbb{R}$ with $0 \le r < 1$, the set $X_r = \{r + n \mid n \in \mathbb{Z}\}$ is an equivalence class of R. To see this, first note that if $n, m \in \mathbb{Z}$, then $(r + n) - (r + m) = n - m \in \mathbb{Z}$. So, any two elements of X_r are equivalent. Also, if x is equivalent to $r + n$, then there is an integer m so that $(r + n) - x = m$. It follows that $x = r + (n - m)$, and so, $x \in X_r$.

Now, if $x \in \mathbb{R}$, then let n be an integer with $n \le x < n + 1$. Then $0 \le x - n < 1$. Let $r = x - n$. We have $x - r = x - (x - n) = n$, and so, $x \in X_r$.

Finally, if $0 \le r < 1$ and $0 \le s < 1$ with $r \le s$ and rRs, then we have $s - r \ge 0$ and $s - r < 1$. Therefore, $s - r = 0$, and so, $s = r$. $\quad\square$

10. Let R be a relation on a set A. Determine if each of the following statements is true or false. If true, provide a proof. If false, provide a counterexample.

   (i)   If R is symmetric and transitive on A, then R is reflexive on A.

   (ii) If R is antisymmetric on A, then R is not symmetric on A.

Solutions:

   (i)   This is **false**. Let $A = \{0, 1\}$ and $R = \{(0, 0)\}$. Then R is symmetric and transitive, but not reflexive (because $(1, 1) \notin R$).

   (ii) This is **false**. $\emptyset$ is both symmetric and antisymmetric on any set A.

11. For $a, b \in \mathbb{N}$, we will say that a divides b, written $a|b$, if there is a natural number k such that $b = ak$. Notice that $|$ is a binary relation on $\mathbb{N}$. Prove that $(\mathbb{N}, |)$ is a partially ordered set, but it is not a linearly ordered set.

Proof: If $a \in \mathbb{N}$ then $a = a \cdot 1$, so that $a|a$. Therefore, $|$ is reflexive. If $a|b$ and $b|a$, then there are natural numbers j and k such that $b = ja$ and $a = kb$. If $a = 0$, then $b = j \cdot 0 = 0$, and so, $a = b$. Suppose $a \neq 0$. We have $a = k(ja) = (kj)a$. Thus, $(kj - 1)a = (kj)a - 1a = 0$. So, $kj - 1 = 0$, and therefore, $kj = 1$. So, $k = j = 1$. Thus, $b = ja = 1a = a$. Therefore, $|$ is antisymmetric. If $a|b$ and $b|c$, then there are natural numbers j and k such that $b = ja$ and $c = kb$. Then $c = kb = k(ja) = (kj)a$. Since the product of two natural numbers is a natural number, $kj \in \mathbb{N}$. So, $a|c$. Therefore, $|$ is transitive. Since $|$ is reflexive, antisymmetric, and transitive on $\mathbb{N}$, $(\mathbb{N}, |)$ is a partially ordered set. Since 2 and 3 do not divide each other, $(\mathbb{N}, |)$ is **not** linearly ordered. □

12. Let P be a partition of a set S. Prove that there is an equivalence relation $\sim$ on S for which the elements of P are the equivalence classes of $\sim$. Conversely, if $\sim$ is an equivalence relation on a set S, prove that the equivalence classes of $\sim$ form a partition of S.

Proof: Let P be a partition of S, and define the relation $\sim$ by $x \sim y$ if and only if there is $X \in P$ with $x, y \in X$.

Let $x \in S$. Since P is a partition of S, $S = \bigcup P$. So, there is $X \in P$ with $x \in X$. It follows that $x \sim x$. Therefore, $\sim$ is reflexive.

If $x \sim y$, then there is $X \in P$ with $x, y \in X$. So, $y, x \in X$ (obviously!). Thus, $y \sim x$, and therefore, $\sim$ is symmetric.

If $x \sim y$ and $y \sim z$, then there are $X, Y \in P$ with $x, y \in X$ and $y, z \in Y$. Since $y \in X$ and $y \in Y$, we have $y \in X \cap Y$. Since P is a partition and $X \cap Y \neq \emptyset$, we must have $X = Y$. So, $z \in X$. Thus, $x, z \in X$, and therefore, $x \sim z$. So, $\sim$ is transitive.

Since $\sim$ is reflexive, symmetric, and transitive on S, $\sim$ is an equivalence relation on S.

We still need to show that $P = \{[x] \mid x \in S\}$. Let $X \in P$ and let $x \in X$. We show that $X = [x]$. Let $y \in X$. Since $x, y \in X$, $x \sim y$. So $y \in [x]$. Thus, $X \subseteq [x]$. Now, let $y \in [x]$. Then $x \sim y$. So, there is $Y \in P$ such that $x, y \in Y$. Since $x \in X$ and $x \in Y$, $x \in X \cap Y$. Since P is a partition and $X \cap Y \neq \emptyset$, we must have $X = Y$. So, $y \in X$. Thus, $[x] \subseteq X$. Since $X \subseteq [x]$ and $[x] \subseteq X$, we have $X = [x]$. Since $X \in P$ was arbitrary, we have shown $P \subseteq \{[x] \mid x \in S\}$.

Now, let $X \in \{[x] \mid x \in S\}$. Then there is $x \in S$ such that $X = [x]$. Since P is a partition of S, $S = \bigcup P$. So, there is $Y \in P$ with $x \in Y$. We will show that $X = Y$. Let $y \in X$. Then $x \sim y$. So, there is $Z \in P$ with $x, y \in Z$. Since $x \in Y$ and $x \in Z$, $x \in Y \cap Z$. Since P is a partition and $Y \cap Z \neq \emptyset$, we must have $Y = Z$. So, $y \in Y$. Since $y \in X$ was arbitrary, $X \subseteq Y$. Now, let $y \in Y$. Then $x \sim y$. So, $y \in [x] = X$. Since $y \in Y$ was arbitrary, $Y \subseteq X$. Since $X \subseteq Y$ and $Y \subseteq X$, we have $X = Y$. Therefore, $X \in P$. Since $X \in \{[x] \mid x \in S\}$ was arbitrary, we have $\{[x] \mid x \in S\} \subseteq P$.

Since $P \subseteq \{[x] \mid x \in S\}$ and $\{[x] \mid x \in S\} \subseteq P$, we have $P = \{[x] \mid x \in S\}$, as desired.

Now, let $\sim$ be an equivalence relation on S. We first show that $\bigcup\{[x] \mid x \in S\} = S$.

Let $y \in \bigcup\{[x] \mid x \in S\}$. Then there is $x \in S$ with $y \in [x]$. By definition of $[x]$, $y \in S$. Therefore, $\bigcup\{[x] \mid x \in S\} \subseteq S$. Now, let $y \in S$. Since $\sim$ is an equivalence relation, $y \sim y$. So, $y \in [y]$. Thus, $y \in \bigcup\{[x] \mid x \in S\}$. So, we have $S \subseteq \bigcup\{[x] \mid x \in S\}$. Since $\bigcup\{[x] \mid x \in S\} \subseteq S$ and $S \subseteq \bigcup\{[x] \mid x \in S\}$, $\bigcup\{[x] \mid x \in S\} = S$.

We next show that if $x, y \in S$, then $[x] \cap [y] = \emptyset$ or $[x] = [y]$.

Suppose $[x] \cap [y] \neq \emptyset$ and let $z \in [x] \cap [y]$. Then $x \sim z$ and $y \sim z$. Since $\sim$ is symmetric, $z \sim y$. Since $\sim$ is transitive, $x \sim y$. By Theorem 2.1.4, $[x] = [y]$.

Since $\bigcup\{[x] \mid x \in S\} = S$ and every pair of equivalence classes are either disjoint or equal, the set of equivalence classes partitions S. $\qquad\square$

Problem Set 3

LEVEL 1

1. Determine if each of the following relations are functions. For each such function, determine if it is injective. State the domain and range of each function.

    (i) $R = \{(a, b), (b, b), (c, d), (e, a)\}$

    (ii) $S = \{(a, a), (a, b), (b, a)\}$

    (iii) $T = \{(a, b) \mid a, b \in \mathbb{R} \wedge b < 0 \wedge a^2 + b^2 = 9\}$

Solutions:

(i) R is a function. It is **not** injective. dom $R = \{a, b, c, e\}$ and ran $R = \{a, b, d\}$.

(ii) S **is not** a function.

(iii) T is a function. It is **not** injective. dom $T = (-3, 3)$ and ran $T = [-3, 0)$.

2. Define $f: \mathbb{Z} \to \mathbb{Z}$ by $f(n) = n^2$. Let $A = \{0, 1, 2, 3, 4\}$, $B = \mathbb{N}$, and $C = \{-2n \mid n \in \mathbb{N}\}$. Evaluate each of the following:

    (i) $f[A]$

    (ii) $f^{-1}[A]$

    (iii) $f^{-1}[B]$

    (iv) $f^{-1}[B \cup C]$

Solutions:

(i) $f[A] = \{0, 1, 4, 9, 16\}$.

(ii) $f^{-1}[A] = \{-2, -1, 0, 1, 2\}$.

(iii) $f^{-1}[B] = \mathbb{Z}$.

(iv) $f^{-1}[B \cup C] = \mathbb{Z}$.

3. In Problems 10 and 11 below, you will be asked to show that $W\left(\frac{\pi}{3}\right) = \left(\frac{1}{2}, \frac{\sqrt{3}}{2}\right)$ and $W\left(\frac{\pi}{6}\right) = \left(\frac{\sqrt{3}}{2}, \frac{1}{2}\right)$. Use this information to compute the sine, cosine, and tangent of each of the following angles:

   (i) $\frac{\pi}{6}$

   (ii) $\frac{\pi}{3}$

   (iii) $\frac{2\pi}{3}$

   (iv) $\frac{5\pi}{6}$

   (v) $\frac{7\pi}{6}$

   (vi) $\frac{4\pi}{3}$

   (vii) $\frac{5\pi}{3}$

   (viii) $\frac{11\pi}{6}$

Solutions:

(i)   By Problem 11, $W\left(\frac{\pi}{6}\right) = \left(\frac{\sqrt{3}}{2}, \frac{1}{2}\right)$. So, $\cos\frac{\pi}{6} = \frac{\sqrt{3}}{2}$, $\sin\frac{\pi}{6} = \frac{1}{2}$, and $\tan\frac{\pi}{6} = \frac{\sin\frac{\pi}{6}}{\cos\frac{\pi}{6}} = \frac{1}{2} \cdot \frac{2}{\sqrt{3}} = \frac{1}{\sqrt{3}}$.

(ii) By Problem 10, $W\left(\frac{\pi}{3}\right) = \left(\frac{1}{2}, \frac{\sqrt{3}}{2}\right)$. So, $\cos\frac{\pi}{3} = \frac{1}{2}$, $\sin\frac{\pi}{3} = \frac{\sqrt{3}}{2}$, and $\tan\frac{\pi}{3} = \frac{\sin\frac{\pi}{3}}{\cos\frac{\pi}{3}} = \frac{\sqrt{3}}{2} \cdot \frac{2}{1} = \mathbf{\sqrt{3}}.$

(iii) Since $\frac{2\pi}{3} = \pi - \frac{\pi}{3}$, by the symmetry of the unit circle, $W\left(\frac{2\pi}{3}\right) = \left(-\frac{1}{2}, \frac{\sqrt{3}}{2}\right)$. It follows that $\cos\frac{2\pi}{3} = -\frac{1}{2}$, $\sin\frac{2\pi}{3} = \frac{\sqrt{3}}{2}$, and $\tan\frac{2\pi}{3} = \frac{\sin\frac{2\pi}{3}}{\cos\frac{2\pi}{3}} = \frac{\sqrt{3}}{2}\left(\frac{-2}{1}\right) = -\sqrt{3}.$

(iv) Since $\frac{5\pi}{6} = \pi - \frac{\pi}{6}$, by the symmetry of the unit circle, $W\left(\frac{5\pi}{6}\right) = \left(-\frac{\sqrt{3}}{2}, \frac{1}{2}\right)$. It follows that $\cos\frac{5\pi}{6} = -\frac{\sqrt{3}}{2}$, $\sin\frac{5\pi}{6} = \frac{1}{2}$, and $\tan\frac{5\pi}{6} = \frac{\sin\frac{5\pi}{6}}{\cos\frac{5\pi}{6}} = \frac{1}{2}\left(-\frac{2}{\sqrt{3}}\right) = -\frac{1}{\sqrt{3}}.$

(v)   Since $\frac{7\pi}{6} = \pi + \frac{\pi}{6}$, by the symmetry of the unit circle, $W\left(\frac{7\pi}{6}\right) = \left(-\frac{\sqrt{3}}{2}, -\frac{1}{2}\right)$. It follows that $\cos\frac{7\pi}{6} = -\frac{\sqrt{3}}{2}$, $\sin\frac{7\pi}{6} = -\frac{1}{2}$, and $\tan\frac{7\pi}{6} = \frac{\sin\frac{7\pi}{6}}{\cos\frac{7\pi}{6}} = -\frac{1}{2}\left(-\frac{2}{\sqrt{3}}\right) = \frac{1}{\sqrt{3}}.$

(vi) Since $\frac{4\pi}{3} = \pi + \frac{\pi}{3}$, by the symmetry of the unit circle, $W\left(\frac{4\pi}{3}\right) = \left(-\frac{1}{2}, -\frac{\sqrt{3}}{2}\right)$. It follows that $\cos\frac{4\pi}{3} = -\frac{1}{2}$, $\sin\frac{4\pi}{3} = -\frac{\sqrt{3}}{2}$, and $\tan\frac{4\pi}{3} = \frac{\sin\frac{4\pi}{3}}{\cos\frac{4\pi}{3}} = -\frac{\sqrt{3}}{2}\left(\frac{-2}{1}\right) = \sqrt{3}.$

(vii) Since $\frac{5\pi}{3} = 2\pi - \frac{\pi}{3}$, by the symmetry of the unit circle, $W\left(\frac{5\pi}{3}\right) = \left(\frac{1}{2}, -\frac{\sqrt{3}}{2}\right)$. It follows that

$$\cos\frac{5\pi}{3} = \frac{1}{2}, \ \sin\frac{5\pi}{3} = -\frac{\sqrt{3}}{2}, \text{ and } \tan\frac{5\pi}{3} = \frac{\sin\frac{5\pi}{3}}{\cos\frac{5\pi}{3}} = -\frac{\sqrt{3}}{2}\cdot\frac{2}{1} = -\sqrt{3}.$$

(viii) Since $\frac{11\pi}{6} = 2\pi - \frac{\pi}{6}$, by the symmetry of the unit circle, $W\left(\frac{11\pi}{6}\right) = \left(\frac{\sqrt{3}}{2}, -\frac{1}{2}\right)$. It follows that

$$\cos\frac{11\pi}{6} = \frac{\sqrt{3}}{2}, \ \sin\frac{11\pi}{6} = -\frac{1}{2}, \text{ and } \tan\frac{11\pi}{6} = \frac{\sin\frac{11\pi}{6}}{\cos\frac{11\pi}{6}} = -\frac{1}{2}\cdot\frac{2}{\sqrt{3}} = -\frac{1}{\sqrt{3}}.$$

LEVEL 2

4. Find sets A and B and a function f such that $f[A \cap B] \neq f[A] \cap f[B]$.

Solution: Define $f: \{a, b\} \to \{0\}$ by $\{(a, 0), (b, 0)\}$. Let $A = \{a\}$ and $B = \{b\}$. Then $A \cap B = \emptyset$. Therefore, $f[A \cap B] = \emptyset$ and $f[A] \cap f[B] = \{0\} \cap \{0\} = \{0\}$.

5. Let $f: A \to B$ and let $V \subseteq B$. Prove that $f[f^{-1}[V]] \subseteq V$.

Proof: Let $y \in f[f^{-1}[V]]$. Then there is $x \in f^{-1}[V]$ with $y = f(x)$. Since $x \in f^{-1}[V]$, we have $y = f(x) \in V$. Since $y \in f[f^{-1}[V]]$ was arbitrary, $f[f^{-1}[V]] \subseteq V$.

6. Use the sum identities (Theorem 3.25) to compute the cosine, sine, and tangent of each of the following angles:

(i) $\frac{5\pi}{12}$

(ii) $\frac{\pi}{12}$

(iii) $\frac{11\pi}{12}$

(iv) $\frac{19\pi}{12}$

Solutions:

(i) $\cos\frac{5\pi}{12} = \cos\left(\frac{\pi}{4} + \frac{\pi}{6}\right) = \cos\frac{\pi}{4}\cos\frac{\pi}{6} - \sin\frac{\pi}{4}\sin\frac{\pi}{6} = \frac{\sqrt{2}}{2}\cdot\frac{\sqrt{3}}{2} - \frac{\sqrt{2}}{2}\cdot\frac{1}{2} = \frac{\sqrt{6} - \sqrt{2}}{4}.$

$\sin\frac{5\pi}{12} = \sin\left(\frac{\pi}{4} + \frac{\pi}{6}\right) = \sin\frac{\pi}{4}\cos\frac{\pi}{6} + \cos\frac{\pi}{4}\sin\frac{\pi}{6} = \frac{\sqrt{2}}{2}\cdot\frac{\sqrt{3}}{2} + \frac{\sqrt{2}}{2}\cdot\frac{1}{2} = \frac{\sqrt{6} + \sqrt{2}}{4}.$

$\tan\frac{5\pi}{12} = \frac{\sin\frac{5\pi}{12}}{\cos\frac{5\pi}{12}} = \frac{\sqrt{6} + \sqrt{2}}{4}\cdot\frac{4}{\sqrt{6} - \sqrt{2}} = \frac{\sqrt{6} + \sqrt{2}}{\sqrt{6} - \sqrt{2}}.$

(ii) $\cos\frac{\pi}{12} = \cos\left(\frac{\pi}{4} - \frac{\pi}{6}\right) = \cos\frac{\pi}{4}\cos\frac{\pi}{6} + \sin\frac{\pi}{4}\sin\frac{\pi}{6} = \frac{\sqrt{2}}{2}\cdot\frac{\sqrt{3}}{2} + \frac{\sqrt{2}}{2}\cdot\frac{1}{2} = \frac{\sqrt{6} + \sqrt{2}}{4}.$

$\sin\frac{\pi}{12} = \sin\left(\frac{\pi}{4} - \frac{\pi}{6}\right) = \sin\frac{\pi}{4}\cos\frac{\pi}{6} - \cos\frac{\pi}{4}\sin\frac{\pi}{6} = \frac{\sqrt{2}}{2}\cdot\frac{\sqrt{3}}{2} - \frac{\sqrt{2}}{2}\cdot\frac{1}{2} = \frac{\sqrt{6} - \sqrt{2}}{4}.$

$\tan\frac{\pi}{12} = \frac{\sin\frac{\pi}{12}}{\cos\frac{\pi}{12}} = \frac{\sqrt{6} - \sqrt{2}}{4}\cdot\frac{4}{\sqrt{6} + \sqrt{2}} = \frac{\sqrt{6} - \sqrt{2}}{\sqrt{6} + \sqrt{2}}.$

(iii) $\cos\dfrac{11\pi}{12} = \cos\left(\dfrac{\pi}{4} + \dfrac{2\pi}{3}\right) = \cos\dfrac{\pi}{4}\cos\dfrac{2\pi}{3} - \sin\dfrac{\pi}{4}\sin\dfrac{2\pi}{3} = \dfrac{\sqrt{2}}{2}\cdot\left(-\dfrac{1}{2}\right) - \dfrac{\sqrt{2}}{2}\cdot\dfrac{\sqrt{3}}{2} = \dfrac{-\sqrt{2}-\sqrt{6}}{4}.$

$\sin\dfrac{11\pi}{12} = \sin\left(\dfrac{\pi}{4} + \dfrac{2\pi}{3}\right) = \sin\dfrac{\pi}{4}\cos\dfrac{2\pi}{3} + \cos\dfrac{\pi}{4}\sin\dfrac{2\pi}{3} = \dfrac{\sqrt{2}}{2}\cdot\left(-\dfrac{1}{2}\right) + \dfrac{\sqrt{2}}{2}\cdot\dfrac{\sqrt{3}}{2} = \dfrac{-\sqrt{2}+\sqrt{6}}{4}.$

$\tan\dfrac{11\pi}{12} = \dfrac{\sin\frac{11\pi}{12}}{\cos\frac{11\pi}{12}} = \dfrac{-\sqrt{2}+\sqrt{6}}{4}\cdot\dfrac{4}{-\sqrt{2}-\sqrt{6}} = \dfrac{-\sqrt{2}+\sqrt{6}}{-\sqrt{2}-\sqrt{6}} = \dfrac{\sqrt{2}-\sqrt{6}}{\sqrt{2}+\sqrt{6}}.$

(iv) $\cos\dfrac{19\pi}{12} = \cos\left(\dfrac{5\pi}{4} + \dfrac{\pi}{3}\right) = \cos\dfrac{5\pi}{4}\cos\dfrac{\pi}{3} - \sin\dfrac{5\pi}{4}\sin\dfrac{\pi}{3} = -\dfrac{\sqrt{2}}{2}\cdot\dfrac{1}{2} - \left(-\dfrac{\sqrt{2}}{2}\right)\cdot\dfrac{\sqrt{3}}{2} = \dfrac{-\sqrt{2}+\sqrt{6}}{4}.$

$\sin\dfrac{19\pi}{12} = \sin\left(\dfrac{5\pi}{4} + \dfrac{\pi}{3}\right) = \sin\dfrac{5\pi}{4}\cos\dfrac{\pi}{3} + \cos\dfrac{5\pi}{4}\sin\dfrac{\pi}{3} = -\dfrac{\sqrt{2}}{2}\cdot\dfrac{1}{2} + \left(-\dfrac{\sqrt{2}}{2}\right)\cdot\dfrac{\sqrt{3}}{2} = \dfrac{-\sqrt{2}-\sqrt{6}}{4}.$

$\tan\dfrac{19\pi}{12} = \dfrac{\sin\frac{19\pi}{12}}{\cos\frac{19\pi}{12}} = \dfrac{-\sqrt{2}-\sqrt{6}}{4}\cdot\dfrac{4}{-\sqrt{2}+\sqrt{6}} = \dfrac{-\sqrt{2}-\sqrt{6}}{-\sqrt{2}+\sqrt{6}} = \dfrac{\sqrt{2}+\sqrt{6}}{\sqrt{2}-\sqrt{6}}.$

LEVEL 3

7. For $f, g \in {}^{\mathbb{R}}\mathbb{R}$, define $f \preccurlyeq g$ if and only if for all $x \in \mathbb{R}$, $f(x) \leq g(x)$. Is $({}^{\mathbb{R}}\mathbb{R}, \preccurlyeq)$ a poset? Is it a linearly ordered set? What if we replace $\preccurlyeq$ by $\preccurlyeq^*$, where $f \preccurlyeq^* g$ if and only if there is an $x \in \mathbb{R}$ such that $f(x) \leq g(x)$?

Solution: If $f \in {}^{\mathbb{R}}\mathbb{R}$, then for all $x \in \mathbb{R}$, $f(x) = f(x)$. So, $f \preccurlyeq f$, and therefore, $\preccurlyeq$ is reflexive.

Let $f, g \in {}^{\mathbb{R}}\mathbb{R}$ with $f \preccurlyeq g$ and $g \preccurlyeq f$. Then for all $x \in \mathbb{R}$, $f(x) \leq g(x)$ and $g(x) \leq f(x)$. So, $f = g$, and therefore, $\preccurlyeq$ is antisymmetric.

Let $f, g, h \in {}^{\mathbb{R}}\mathbb{R}$ with $f \preccurlyeq g$ and $g \preccurlyeq h$. Then for all $x \in \mathbb{R}$, $f(x) \leq g(x)$ and $g(x) \leq h(x)$. So, by the transitivity of $\leq$, for all $x \in \mathbb{R}$, $f(x) \leq h(x)$. Thus, $f \preccurlyeq h$, and therefore, $\preccurlyeq$ is transitive.

Since $\preccurlyeq$ is reflexive, antisymmetric, and transitive, $({}^{\mathbb{R}}\mathbb{R}, \preccurlyeq)$ is a poset.

Let $f(x) = x$ and $g(x) = x^2$. Then $f(2) = 2$ and $g(2) = 4$. So, $f(2) < g(2)$. Therefore, $g \not\preccurlyeq f$. We also have $f\left(\frac{1}{2}\right) = \frac{1}{2}$ and $g\left(\frac{1}{2}\right) = \frac{1}{4}$. So, $g\left(\frac{1}{2}\right) < f\left(\frac{1}{2}\right)$. Therefore, $f \not\preccurlyeq g$. So, f and g are incomparible with respect to $\preccurlyeq$. Therefore, $({}^{\mathbb{R}}\mathbb{R}, \preccurlyeq)$ is **not** a linearly ordered set.

The same example from the last paragraph gives us $f \preccurlyeq^* g$ and $g \preccurlyeq^* f$. But $f \neq g$. So, $\preccurlyeq^*$ is **not** antisymmetric, and therefore, $({}^{\mathbb{R}}\mathbb{R}, \preccurlyeq^*)$ is **not** a poset.

8. Prove that the function $f: \mathbb{N} \to \mathbb{Z}$ defined by $f(n) = \begin{cases} \dfrac{n}{2} & \text{if } n \text{ is even} \\ -\dfrac{n+1}{2} & \text{if } n \text{ is odd} \end{cases}$ is a bijection.

Proof: First note that if n is even, then there is $k \in \mathbb{Z}$ with $n = 2k$, and so, $\frac{n}{2} = \frac{2k}{2} = k \in \mathbb{Z}$, and if n is odd, there is $k \in \mathbb{Z}$ with $n = 2k + 1$, and so, $-\frac{n+1}{2} = -\frac{(2k+1)+1}{2} = -\frac{2k+2}{2} = -\frac{2(k+1)}{2} = -(k+1) \in \mathbb{Z}$. So, f does take each natural number to an integer.

Now, suppose that $n, m \in \mathbb{N}$ with $f(n) = f(m)$. If n and m are both even, we have $\frac{n}{2} = \frac{m}{2}$, and so, $2 \cdot \frac{n}{2} = 2 \cdot \frac{m}{2}$. Thus, $n = m$. If n and m are both odd, we have $-\frac{n+1}{2} = -\frac{m+1}{2}$, and so, $\frac{n+1}{2} = \frac{m+1}{2}$. Thus, $2 \cdot \frac{n+1}{2} = 2 \cdot \frac{m+1}{2}$. So, $n + 1 = m + 1$, and therefore, $n = m$. If n is even and m is odd, then we have $\frac{n}{2} = -\frac{m+1}{2}$. So, $2 \cdot \frac{n}{2} = 2\left(-\frac{m+1}{2}\right)$. Therefore, $n = -(m + 1)$. Since $m \in \mathbb{N}$, $m \geq 0$. So, $m + 1 \geq 1$. Therefore, $n = -(m + 1) \leq -1$, contradicting $n \in \mathbb{N}$. So, it is impossible for n to be even, m to be odd, and $f(n) = f(m)$. Similarly, we cannot have n odd and m even. So, f is an injection.

Now, let $k \in \mathbb{Z}$. If $k \geq 0$, then $2k \in \mathbb{N}$ and $f(2k) = \frac{2k}{2} = k$. If $k < 0$, then $-2k > 0$, and so, we have $-2k - 1 \in \mathbb{N}$. Then $f(-2k - 1) = -\frac{(-2k-1)+1}{2} = -\frac{-2k}{2} = k$. So, f is a surjection.

Since f is both an injection and a surjection, f is a bijection.

---

9. Define $f : \mathbb{R} \setminus \{2\} \to \mathbb{R}$ by $f(x) = \frac{3x}{x-2}$.

  (i)   Prove that f is injective.

  (ii) Prove that f is **not** surjective.

  (iii) Find a set X such that $f : \mathbb{R} \setminus \{2\} \to X$ is a bijection.

  (iv) What is f^{-1}?

Proofs:

  (i)   Suppose that $a, b \in \mathbb{R} \setminus \{2\}$ and $f(a) = f(b)$. Then $\frac{3a}{a-2} = \frac{3b}{b-2}$. It follows that $3a(b - 2) = 3b(a - 2)$, and so, $3ab - 6a = 3ab - 6b$. So, $-6a = -6b$. Thus, $a = b$. □

  (ii) Let's show that $3 \notin \operatorname{ran} f$. If $\frac{3x}{x-2} = 3$, then $3x = 3(x - 2) = 3x - 6$, and therefore, we have $0 = -6$, which is false. Since $3 \notin \operatorname{ran} f$, f is not surjective. □

  (iii) If $y = \frac{3x}{x-2}$, then $y(x - 2) = 3x$. So, $xy - 2y = 3x$, and therefore, $xy - 3x = 2y$. So, we have $x(y - 3) = 2y$, and so, $x = \frac{2y}{y-3}$. Interchanging the roles of x and y shows that the inverse of f is $f^{-1}(x) = \frac{2x}{x-3}$.

   So, if $b \neq 3$, we have $f\left(\frac{2b}{b-3}\right) = \frac{3\left(\frac{2b}{b-3}\right)}{\frac{2b}{b-3}-2} = \frac{6b}{2b-2(b-3)} = \frac{6b}{2b-2b+6} = \frac{6b}{6} = b$. It follows that $b \in \operatorname{ran} f$. So, if we let $X = \mathbb{R} \setminus \{3\}$, then $f : \mathbb{R} \setminus \{2\} \to X$ is a bijection.

  (iv) We found in part (iii) above that $f^{-1} : \mathbb{R} \setminus \{3\} \to \mathbb{R} \setminus \{2\}$ is defined by $f^{-1}(x) = \frac{2x}{x-3}$.

10. Consider triangle AOP, where $O = (0,0)$, $A = (1,0)$, and P is the point on the unit circle so that angle POA has radian measure $\frac{\pi}{3}$. Prove that triangle AOP is equilateral, and then use this to prove that $W\left(\frac{\pi}{3}\right) = \left(\frac{1}{2}, \frac{\sqrt{3}}{2}\right)$. You may use the following facts about triangles: (i) The interior angle measures of a triangle sum to π radians; (ii) Two sides of a triangle have the same length if and only if the interior angles of the triangle opposite these sides have the same measure; (iii) If two sides of a triangle have the same length, then the line segment beginning at the point of intersection of those two sides and terminating on the opposite base midway between the endpoints of that base is perpendicular to that base.

Proof: Let's start by drawing the unit circle together with triangle AOP. We also draw line segment PE, where E is midway between O and A.

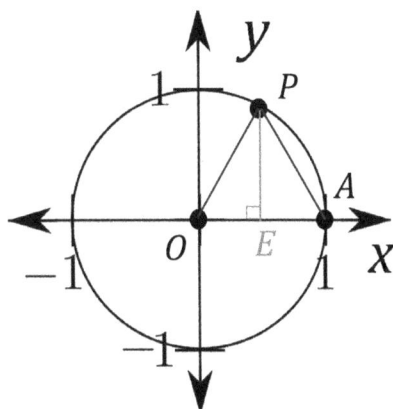

Since OP and OA are both radii of the circle, they have the same length. By (ii), angles OAP and OPA have the same measure. By (i), the sum of these measures is $\pi - \frac{\pi}{3} = \frac{3\pi}{3} - \frac{\pi}{3} = \frac{2\pi}{3}$. So, each of angles OAP and OPA measure $\frac{\pi}{3}$ radians. It follows from (ii) again that triangle AOP is equilateral.

By (iii), PE is perpendicular to OA.

Now, $OP = 1$ because OP is a radius of the unit circle and $OE = \frac{1}{2}$ because OA is a radius of the unit circle and E is midway between O and A. Since triangle OEP is a right triangle with hypotenuse OP, by the Pythagorean Theorem, $PE^2 = OP^2 - OE^2 = 1^2 - \left(\frac{1}{2}\right)^2 = 1 - \frac{1}{4} = \frac{3}{4}$. So, $PE = \sqrt{\frac{3}{4}} = \frac{\sqrt{3}}{\sqrt{4}} = \frac{\sqrt{3}}{2}$. It follows that $W\left(\frac{\pi}{3}\right) = \left(\frac{1}{2}, \frac{\sqrt{3}}{2}\right)$. $\qquad\square$

11. Prove that $W\left(\frac{\pi}{6}\right) = \left(\frac{\sqrt{3}}{2}, \frac{1}{2}\right)$. You can use facts (i), (ii), and (iii) described in Problem 10.

Proof: Let's start by drawing a picture similar to what we drew in Problem 10. We draw P and Q on the unit circle and A on the positive x-axis so that angle AOP has radian measure $\frac{\pi}{6}$, angle AOQ has radian measure $-\frac{\pi}{6}$, and A is right in the middle of the line segment joining P and Q.

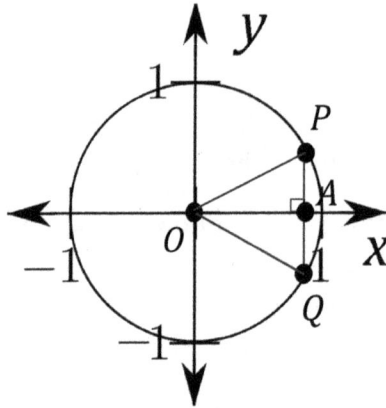

By reasoning similar to what was done in Problem 10, we see that triangle POQ is equilateral and OA is perpendicular to PQ.

Now, $OP = 1$ because OP is a radius of the unit circle and $AP = \frac{1}{2}$ because A is midway between P and Q. Since triangle POA is a right triangle with hypotenuse OP, by the Pythagorean Theorem, $OA^2 = OP^2 - AP^2 = 1^2 - \left(\frac{1}{2}\right)^2 = 1 - \frac{1}{4} = \frac{3}{4}$. Therefore, $OA = \sqrt{\frac{3}{4}} = \frac{\sqrt{3}}{\sqrt{4}} = \frac{\sqrt{3}}{2}$. It follows that $W\left(\frac{\pi}{6}\right) = \left(\frac{\sqrt{3}}{2}, \frac{1}{2}\right)$. □

12. Let θ and ϕ be the radian measure of angles A and B, respectively. Prove the following identity:
$$\cos(\theta - \phi) = \cos\theta\cos\phi + \sin\theta\sin\phi$$

Proof: Let's draw a picture of the unit circle together with angles θ, ϕ, and $\theta - \phi$ in standard position, and label the corresponding points on the unit circle.

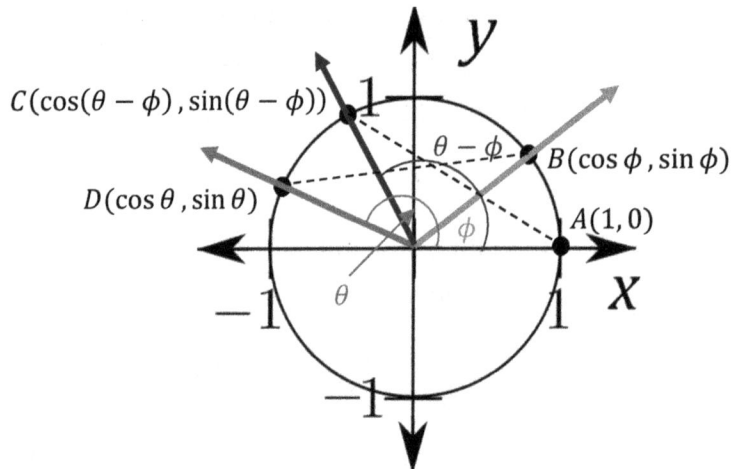

Since the arcs moving counterclockwise from A to C and from B to D both have radian measure $\theta - \phi$, it follows that $AC = BD$, and so, using the Pythagorean Theorem twice, we have
$$(\cos(\theta - \phi) - 1)^2 + (\sin(\theta - \phi) - 0)^2 = (\cos\theta - \cos\phi)^2 + (\sin\theta - \sin\phi)^2$$

32

The left-hand side of this equation is equal to:

$$(\cos(\theta - \phi) - 1)^2 + (\sin(\theta - \phi) - 0)^2$$
$$= \cos^2(\theta - \phi) - 2\cos(\theta - \phi) + 1 + \sin^2(\theta - \phi)$$
$$= (\cos^2(\theta - \phi) + \sin^2(\theta - \phi)) - 2\cos(\theta - \phi) + 1$$
$$= 1 - 2\cos(\theta - \phi) + 1 \text{ (by the Pythagorean Identity)}$$
$$= 2 - 2\cos(\theta - \phi)$$

The right-hand side of this equation is equal to:

$$(\cos\theta - \cos\phi)^2 + (\sin\theta - \sin\phi)^2$$
$$= \cos^2\theta - 2\cos\theta\cos\phi + \cos^2\phi + \sin^2\theta - 2\sin\theta\sin\phi + \sin^2\phi$$
$$= (\cos^2\theta + \sin^2\theta) + (\cos^2\phi + \sin^2\phi) - 2\cos\theta\cos\phi - 2\sin\theta\sin\phi$$
$$= 1 + 1 - 2\cos\theta\cos\phi - 2\sin\theta\sin\phi$$
$$= 2 - 2\cos\theta\cos\phi - 2\sin\theta\sin\phi$$

Therefore, we have $2 - 2\cos(\theta - \phi) = 2 - 2\cos\theta\cos\phi - 2\sin\theta\sin\phi$. Subtracting 2 from each side of this equation gives us $-2\cos(\theta - \phi) = -2\cos\theta\cos\phi - 2\sin\theta\sin\phi$. Multiplying each side of this last equation by $-\frac{1}{2}$ gives us $\cos(\theta - \phi) = \cos\theta\cos\phi + \sin\theta\sin\phi$, as desired. □

13. Let θ and ϕ be the radian measure of angles A and B, respectively. Prove the following identities:

    (i) $\cos(\theta + \phi) = \cos\theta\cos\phi - \sin\theta\sin\phi$

    (ii) $\cos(\pi - \theta) = -\cos\theta$

    (iii) $\cos\left(\frac{\pi}{2} - \theta\right) = \sin\theta$

    (iv) $\sin\left(\frac{\pi}{2} - \theta\right) = \cos\theta$

    (v) $\sin(\theta + \phi) = \sin\theta\cos\phi + \cos\theta\sin\phi$

    (vi) $\sin(\pi - \theta) = \sin\theta$

Proofs:

    (i) $\cos(\theta + \phi) = \cos(\theta - (-\phi)) = \cos\theta\cos(-\phi) + \sin\theta\sin(-\phi)$ (by Problem 12)

                     $= \cos\theta\cos\phi - \sin\theta\sin\phi$ (by the Negative Identities). □

    (ii) $\cos(\pi - \theta) = \cos\pi\cos\theta + \sin\pi\sin\theta = (-1)\cos\theta + 0\cdot\sin\theta = -\cos\theta$. □

    (iii) $\cos\left(\frac{\pi}{2} - \theta\right) = \cos\frac{\pi}{2}\cos\theta + \sin\frac{\pi}{2}\sin\theta = 0\cdot\cos\theta + 1\cdot\sin\theta = \sin\theta$. □

    (iv) $\sin\left(\frac{\pi}{2} - \theta\right) = \cos\left(\frac{\pi}{2} - \left(\frac{\pi}{2} - \theta\right)\right) = \cos\left(\frac{\pi}{2} - \frac{\pi}{2} + \theta\right) = \cos\theta$. □

    (v) $\sin(\theta + \phi) = \cos\left(\frac{\pi}{2} - (\theta + \phi)\right) = \cos\left(\left(\frac{\pi}{2} - \theta\right) - \phi\right)$

           $= \cos\left(\frac{\pi}{2} - \theta\right)\cos\phi + \sin\left(\frac{\pi}{2} - \theta\right)\sin\phi = \sin\theta\cos\phi + \cos\theta\sin\phi$. □

(vi) $\sin(\pi - \theta) = \sin\big(\pi + (-\theta)\big) = \sin\pi\cos(-\theta) + \cos\pi\sin(-\theta)$

$$= 0 \cdot \cos\theta + (-1)(-\sin\theta) = \sin\theta. \qquad \square$$

LEVEL 5

14. Let X be a nonempty set of sets and let f be a function such that $\cup X \subseteq \operatorname{dom} f$. Prove each of the following:

 (i) $f[\cup X] = \cup\{f[A] \mid A \in X\}$

 (ii) $f[\cap X] \subseteq \cap\{f[A] \mid A \in X\}$

 (iii) $f^{-1}[\cup X] = \cup\{f^{-1}[A] \mid A \in X\}$

 (iv) $f^{-1}[\cap X] = \cap\{f^{-1}[A] \mid A \in X\}$

Proofs:

(i)   Let $y \in f[\cup X]$. Then there is $x \in \cup X$ such that $f(x) = y$. Since $x \in \cup X$, there is $B \in X$ such that $x \in B$. So, $y = f(x) \in f[B]$. Therefore, $y \in \cup\{f[A] \mid A \in X\}$. Since $y \in f[\cup X]$ was arbitrary, we see that $f[\cup X] \subseteq \cup\{f[A] \mid A \in X\}$.

Now, let $y \in \cup\{f[A] \mid A \in X\}$. Then there is $B \in X$ such that $y \in f[B]$. So, there is $x \in B$ such that $y = f(x)$. By Problem 26 (part (i)) from Problem Set 1, $B \subseteq \cup X$. Since $x \in B$ and $B \subseteq \cup X, x \in \cup X$. Thus, $y = f(x) \in f[\cup X]$. Since $y \in \cup\{f[A] \mid A \in X\}$ was arbitrary, we see that $\cup\{f[A] \mid A \in X\} \subseteq f[\cup X]$.

Since $f[\cup X] \subseteq \cup\{f[A] \mid A \in X\}$ and $\cup\{f[A] \mid A \in X\} \subseteq f[\cup X]$, it follows that $f[\cup X] = \cup\{f[A] \mid A \in X\}$. $\qquad \square$

(ii) Let $y \in f[\cap X]$. Then there is $x \in \cap X$ such that $f(x) = y$. Let $B \in X$. Since $x \in \cap X, x \in B$. So, $y = f(x) \in f[B]$. Since $B \in X$ was arbitrary, $y \in \cap\{f[A] \mid A \in X\}$. Since $y \in f[\cap X]$ was arbitrary, we see that $f[\cap X] \subseteq \cap\{f[A] \mid A \in X\}$. $\qquad \square$

(iii) $x \in f^{-1}[\cup X]$ if and only if $f(x) \in \cup X$ if and only if there is $A \in X$ such that $f(x) \in A$ if and only if there is $A \in X$ such that $x \in f^{-1}[A]$ if and only if $x \in \cup\{f^{-1}[A] \mid A \in X\}$. Therefore, $f^{-1}[\cup X] = \cup\{f^{-1}[A] \mid A \in X\}$. $\qquad \square$

(iv) $x \in f^{-1}[\cap X]$ if and only if $f(x) \in \cap X$ if and only for all $A \in X, f(x) \in A$ if and only if for all $A \in X$, $x \in f^{-1}[A]$ if and only if $x \in \cap\{f^{-1}[A] \mid A \in X\}$. Therefore, we see that $f^{-1}[\cap X] = \cap\{f^{-1}[A] \mid A \in X\}$. $\qquad \square$

15. Given ordered pairs $(x_1, y_1), (x_2, y_2)$, with x_1 and x_2 distinct, prove that there is a unique linear function f such that $f(x_1) = y_1$ and $f(x_2) = y_2$.

Proof: Let $m = \frac{y_2 - y_1}{x_2 - x_1}$, let $b = y_1 - mx_1$, and define $f: \mathbb{R} \to \mathbb{R}$ by $f(x) = mx + b$. Then f is a linear function and we have

$$f(x_1) = mx_1 + b = mx_1 + y_1 - mx_1 = y_1,$$

$$f(x_2) = mx_2 + b = mx_2 + y_1 - mx_1 = m(x_2 - x_1) + y_1 = \frac{y_2 - y_1}{x_2 - x_1}(x_2 - x_1) + y_1 = y_2 - y_1 + y_1 = y_2.$$

So, such a linear function exists.

Suppose that $g: \mathbb{R} \to \mathbb{R}$ is an arbitrary linear function such that $g(x_1) = y_1$ and $g(x_2) = y_2$, say $g(x) = kx + d$. Then $y_1 = kx_1 + d$ and $y_2 = kx_2 + d$. Subtracting the first equation from the second yields $y_2 - y_1 = k(x_2 - x_1)$, and so, $k = \frac{y_2 - y_1}{x_2 - x_1} = m$. It follows that $d = y_1 - kx_1 = y_1 - mx_1 = b$. Therefore, $g = f$. So, the linear function f is unique. $\qquad\square$

Problem Set 4

LEVEL 1

1. Use the Principle of Mathematical Induction to prove each of the following:

   (i) $2^n > n$ for all natural numbers $n \geq 1$.

   (ii) $0 + 1 + 2 + \cdots + n = \frac{n(n+1)}{2}$ for all natural numbers.

   (iii) $0^2 + 1^2 + 2^2 + \cdots + n^2 = \frac{n(n+1)(2n+1)}{6}$ for all natural numbers.

   (iv) $n! > 2^n$ for all natural numbers $n \geq 4$ (where $n! = 1 \cdot 2 \cdots n$ for all natural numbers $n \geq 1$).

   (v) $2^n \geq n^2$ for all natural numbers $n \geq 4$.

Proofs:

   (i) **Base Case** $(k = 1)$: $2^1 = 2 > 1$.

Inductive Step: Let $k \in \mathbb{N}$ with $k \geq 1$ and assume that $2^k > k$. Then we have

$$2^{k+1} = 2^k \cdot 2^1 = 2^k \cdot 2 > k \cdot 2 = 2k = k + k \geq k + 1.$$

Therefore, $2^{k+1} > k + 1$.

By the Principle of Mathematical Induction, $2^n > n$ for all natural numbers $n \geq 1$. $\quad\square$

   (ii) **Base Case** $(k = 0)$: $0 = \frac{0(0+1)}{2}$.

Inductive Step: Let $k \in \mathbb{N}$ and assume that $0 + 1 + 2 + \cdots + k = \frac{k(k+1)}{2}$. Then we have

$$0 + 1 + 2 + \cdots + k + (k+1) = \frac{k(k+1)}{2} + (k+1) = (k+1)\left(\frac{k}{2}+1\right) = (k+1)\left(\frac{k}{2}+\frac{2}{2}\right)$$

$$= (k+1)\left(\frac{k+2}{2}\right) = \frac{(k+1)(k+2)}{2} = \frac{(k+1)\big((k+1)+1\big)}{2}$$

By the Principle of Mathematical Induction, $0 + 1 + 2 + \cdots + n = \frac{n(n+1)}{2}$ for all natural numbers n. $\quad\square$

   (iii) **Base Case** $(k = 0)$: $0^2 = 0$ and $\frac{0(0+1)(2 \cdot 0+1)}{6} = 0$.

Inductive Step: Let $k \in \mathbb{N}$ and assume that $0^2 + 1^2 + 2^2 + \cdots + k^2 = \frac{k(k+1)(2k+1)}{6}$. Then we have

$$0^2 + 1^2 + 2^2 + \cdots + k^2 + (k+1)^2 = \frac{k(k+1)(2k+1)}{6} + (k+1)^2$$

$$= (k+1)\left(\frac{k(2k+1)}{6} + (k+1)\right) = (k+1)\left(\frac{2k^2+k}{6} + \frac{6(k+1)}{6}\right) = (k+1)\left(\frac{2k^2+7k+6}{6}\right)$$

$$= (k+1)\left(\frac{(k+2)(2k+3)}{6}\right) = \frac{(k+1)(k+2)(2k+3)}{6} = \frac{(k+1)\big((k+1)+1\big)(2(k+1)+1)}{6}$$

36

By the Principle of Mathematical Induction, $0^2 + 1^2 + 2^2 + \cdots + n^2 = \frac{n(n+1)(2n+1)}{6}$ for all natural numbers n. $\square$

(iv) **Base Case** $(k = 4)$: $4! = 1 \cdot 2 \cdot 3 \cdot 4 = 24 > 16 = 2^4$.

Inductive Step: Let $k \in \mathbb{N}$ with $k \geq 4$ and assume that $k! > 2^k$. Then we have

$$(k+1)! = (k+1)k! > (k+1)2^k \geq (4+1) \cdot 2^k = 5 \cdot 2^k \geq 2 \cdot 2^k = 2^1 \cdot 2^k = 2^{1+k} = 2^{k+1}.$$

Therefore, $(k+1)! > 2^{k+1}$.

By the Principle of Mathematical Induction, $n! > 2^n$ for all natural numbers $n \geq 4$. $\square$

(v) **Base Case** $(k = 4)$: $2^4 = 16 = 4^2$. So, $2^4 \geq 4^2$.

Inductive Step: Let $k \in \mathbb{N}$ with $k \geq 4$ and assume that $2^k \geq k^2$. Then we have

$$2^{k+1} = 2^k \cdot 2^1 \geq k^2 \cdot 2 = 2k^2 = k^2 + k^2.$$

By Theorem 4.14, $k^2 > 2k + 1$. So, we have $2^{k+1} > k^2 + 2k + 1 = (k+1)^2$.

Therefore, $2^{k+1} \geq (k+1)^2$.

By the Principle of Mathematical Induction, $2^n \geq n^2$ for all $n \in \mathbb{N}$ with $n \geq 4$. $\square$

Note: Let's take one last look at number (iv). $2^0 = 1 \geq 0 = 0^2$. So, the statement in (iv) is true for $k = 0$. Also, $2^1 = 2 \geq 1 = 1^2$ and $2^2 = 4 = 2^2$. So, the statement is true for $k = 1$ and $k = 2$. However, $2^3 = 8$ and $3^2 = 9$. So, the statement is false for $k = 3$. It follows that $2^n \geq n^2$ for all natural numbers n except $n = 3$.

2. A natural number n is **divisible** by a natural number k, written $k|n$, if there is another natural number b such that $n = kb$. Prove that $n^3 - n$ is divisible by 3 for all natural numbers n.

Proof by Mathematical Induction:

Base Case $(k = 0)$: $0^3 - 0 = 0 = 3 \cdot 0$. So, $0^3 - 0$ is divisible by 3.

Inductive Step: Let $k \in \mathbb{N}$ and assume that $k^3 - k$ is divisible by 3. Then $k^3 - k = 3b$ for some integer b. Now,

$$(k+1)^3 - (k+1) = (k+1)[(k+1)^2 - 1] = (k+1)[(k+1)(k+1) - 1]$$

$$= (k+1)(k^2 + 2k + 1 - 1) = (k+1)(k^2 + 2k) = k^3 + 2k^2 + k^2 + 2k = k^3 + 3k^2 + 2k$$

$$= k^3 - k + k + 3k^2 + 2k = (k^3 - k) + 3k^2 + 3k = 3b + 3(k^2 + k) = 3(b + k^2 + k).$$

Since $\mathbb{Z}$ is closed under addition and multiplication, $b + k^2 + k \in \mathbb{Z}$. Therefore, $(k+1)^3 - (k+1)$ is divisible by 3.

By the Principle of Mathematical Induction, $n^3 - n$ is divisible by 3 for all $n \in \mathbb{N}$. $\square$

Note: Notice our use of SACT (see Note 7 following the proof of Theorem 4.11) in the beginning of the last line of the sequence of equations. We needed $k^3 - k$ to appear, but the $-k$ was nowhere to be found. So, we simply threw it in, and then repaired the damage by adding k right after it.

3. Let $z = -4 - i$ and $w = 3 - 5i$. Compute each of the following:

   (i) $z + w$

   (ii) zw

   (iii) $\text{Im } w$

Solutions:

   (i) $z + w = (-4 - i) + (3 - 5i) = (-4 + 3) + (-1 - 5)i = \mathbf{-1 - 6i}$.

   (ii) $zw = (-4 - i)(3 - 5i) = (-12 - 5) + (20 - 3)i = \mathbf{-17 + 17i}$.

   (iii) $\text{Im } w = \text{Im } (3 - 5i) = \mathbf{-5}$.

LEVEL 2

4. Prove each of the following. (You may assume that $<$ is a strict linear ordering of $\mathbb{N}$.)

   (i)   Addition is commutative in $\mathbb{N}$.

   (ii)   The set of natural numbers is closed under multiplication.

   (iii) 1 is a multiplicative identity in $\mathbb{N}$.

   (iv)   Multiplication is distributive over addition in $\mathbb{N}$.

   (v)   Multiplication is associative in $\mathbb{N}$.

   (vi)   Multiplication is commutative in $\mathbb{N}$.

   (vii) For all natural numbers m, n, and k, if $m + k = n + k$, then $m = n$.

   (viii) For all natural numbers m, n, and k with $k \neq 0$, if $mk = nk$, then $m = n$.

   (ix)   For all natural numbers m and n, $m < n$ if and only if there is a natural number $k > 0$ such that $n = m + k$.

   (x)   For all natural numbers m, n, and k, $m < n$ if and only if $m + k < n + k$.

   (xi)   For all natural numbers m and n, if $m > 0$ and $n > 0$, then $mn > 0$.

Proofs:

   (i)   We first prove by induction on n that $1 + n = n + 1$.

Base Case ($k = 0$): By definition of addition of natural numbers, we have $1 + 0 = 1$. By Theorem 4.8, we have $0 + 1 = 1$. Therefore, $1 + 0 = 0 + 1$.

Inductive Step: Let $k \in \mathbb{N}$ and assume that $1 + k = k + 1$. Then we have

$$1 + (k + 1) = (1 + k) + 1 = (k + 1) + 1.$$

For the first equality, we used the definition of addition of natural numbers. For the second equality, we used the inductive hypothesis.

By the Principle of Mathematical Induction, for all natural numbers n, $1 + n = n + 1$.

We are now ready to use induction to prove the result. Assume that m is a natural number.

Base Case $(k = 0)$: By definition of addition of natural numbers, $m + 0 = m$. By Theorem 4.8, $0 + m = m$. Therefore, $m + 0 = 0 + m$.

Inductive Step: Let $k \in \mathbb{N}$ and assume that $m + k = k + m$. Then we have

$$m + (k + 1) = (m + k) + 1 = (k + m) + 1 = k + (m + 1) = k + (1 + m) = (k + 1) + m.$$

For the first and third equalities, we used the definition of addition of natural numbers (or Theorem 4.9). For the second equality, we used the inductive hypothesis. For the fourth equality, we used the preliminary result that we proved above. For the fifth equality, we used Theorem 4.9.

By the Principle of Mathematical Induction, for all natural numbers n, $m + n = n + m$.

Since m was an arbitrary natural number, we have shown that for all natural numbers m and n, we have $m + n = n + m$. □

    (ii)   Assume that m is a natural number.

Base Case $(k = 0)$: $m \cdot 0 = 0$, which is a natural number.

Inductive Step: Let k be a natural number and assume that mk is also a natural number. Then $m(k + 1) = mk + m$. Since mk and m are both natural numbers, by Theorem 4.7, $mk + m$ is a natural number.

By the Principle of Mathematical Induction, mn is a natural number for all natural numbers n.

Since m was an arbitrary natural number, we have shown that the product of any two natural numbers is a natural number. □

    (iii)   Assume that m is a natural number.

We have $m \cdot 1 = m(0 + 1) = m \cdot 0 + m = 0 + m = m$. For the first and fourth equalities, we used Theorem 4.8. For the second and third equalities, we used the definition of multiplication of natural numbers.

We prove that $1 \cdot n = n$ by induction on n.

Proof: Base Case $(k = 0)$: $1 \cdot 0 = 0$ by the definition of multiplication of natural numbers.

Inductive Step: Let $k \in \mathbb{N}$ and assume that $1 \cdot k = k$. Then

$$1(k + 1) = 1 \cdot k + 1 = k + 1.$$

For the first equality, we used the definition of multiplication of natural numbers. For the second equality, we used the inductive hypothesis.

By the Principle of Mathematical Induction, for all natural numbers n, $1 \cdot n = n$. □

(iv)   Let m and n be natural numbers. We first prove that for all $t \in \mathbb{N}$, $(m + n) \cdot t = mt + nt$ (we say that multiplication is **right distributive** over addition in $\mathbb{N}$).

Base Case ($k = 0$): $(m + n) \cdot 0 = 0$ by the definition of multiplication of natural numbers. Similarly, $m \cdot 0 = 0$ and $n \cdot 0 = 0$. So, $m \cdot 0 + n \cdot 0 = 0$. Therefore, $(m + n) \cdot 0 = m \cdot 0 + n \cdot 0$.

Inductive Step: Let $k \in \mathbb{N}$ and assume that $(m + n) \cdot k = mk + nk$. Then

$$(m + n)(k + 1) = (m + n) \cdot k + (m + n) = (mk + nk) + (m + n)$$
$$= (mk + m) + (nk + n) = m(k + 1) + n(k + 1).$$

For the first and fourth equalities, we used the definition of multiplication of natural numbers. For the second equality, we used the inductive hypothesis. For the third equality we used the fact that addition is associative and commutative in $\mathbb{N}$ several times.

By the Principle of Mathematical Induction, for all natural numbers t, $(m + n) \cdot t = mt + nt$.

We next prove that $n \cdot 0 = 0$ and $0 \cdot n = 0$ for all natural numbers n.

$n \cdot 0 = 0$ by the definition of multiplication of natural numbers.

We prove that $0 \cdot n = 0$ by induction on n.

Base Case ($k = 0$): By definition of multiplication of natural numbers, we have $0 \cdot 0 = 0$.

Inductive Step: Let $k \in \mathbb{N}$ and assume that $0 \cdot k = 0$. Then we have

$$0 \cdot (k + 1) = 0 \cdot k + 0 = 0 + 0 = 0.$$

For the first equality, we used the definition of multiplication of natural numbers. For the second equality, we used the inductive hypothesis. For the third equality, we used the definition of addition of natural numbers.

By the Principle of Mathematical Induction, for all natural numbers n, $0 \cdot n = n$.

Let m be a natural number. We prove that for all $n \in \mathbb{N}$, $mn = nm$ (we say that multiplication is **commutative** in $\mathbb{N}$).

Base Case ($k = 0$): $m \cdot 0 = 0$ by the definition of multiplication of natural numbers. We just proved that $0 \cdot m = 0$. Therefore, $m \cdot 0 = 0 \cdot m$

Inductive Step: Let $k \in \mathbb{N}$ and assume that $mk = km$. Then

$$m(k + 1) = mk + m = km + m = (k + 1)m.$$

For the first equality, we used the definition of multiplication of natural numbers. For the second equality, we used the inductive hypothesis. For the third equality we used the fact that multiplication is right distributive over addition in $\mathbb{N}$ (proved above).

By the Principle of Mathematical Induction, for all natural numbers n, $mn = nm$.

Finally, let $m, n, t \in \mathbb{N}$. Then $m(n + t) = (n + t)m = nm + tm = mn + mt$. This shows that multiplication is distributive over addition in $\mathbb{N}$. $\square$

(v)   Assume that m and n are natural numbers.

Base Case $(k = 0)$: $(mn) \cdot 0 = 0 = m \cdot 0 = m(n \cdot 0)$ by the definition of multiplication of natural numbers.

Inductive Step: Let $k \in \mathbb{N}$ and assume that $(mn)k = m(nk)$. Then

$$(mn)(k + 1) = (mn)k + mn = m(nk) + mn = mn + m(nk)$$
$$= m(n + nk) = m(nk + n) = m\big(n(k + 1)\big).$$

For the first and sixth equalities, we used the definition of multiplication of natural numbers. For the second equality, we used the inductive hypothesis. For the third and fifth equalities, we used the fact that addition is commutative in $\mathbb{N}$. For the fourth equality, we used the fact that multiplication is distributive over addition in $\mathbb{N}$.

By the Principle of Mathematical Induction, for all natural numbers t, $(mn)t = m(nt)$. This shows that multiplication is associative in $\mathbb{N}$. □

(vi)   This was already proved in (iv) above. □

(vii)   Let $m, n \in \mathbb{N}$. We prove by induction on k that $m + k = n + k \rightarrow m = n$.

Base Case ($k = 0$ and $k = 1$): If $m + 0 = n + 0$, then since $m + 0 = m$ and $n + 0 = n$ (by definition of addition of natural numbers), $m = n$. Next, suppose $m + 1 = n + 1$. Then $m \cup \{m\} = n \cup \{n\}$. If $n \neq m$, then either $n \in m$ or $m \in n$. Without loss of generality, assume that $n \in m$. Since $\in$ is antisymmetric on $\mathbb{N}$ and $n \neq m$, we must have $m \notin n$. Thus, $m = n$, contrary to our assumption that $n \neq m$. This contradiction shows that $m = n$.

Inductive Step: Let $t \in \mathbb{N}$, assume that $m + t = n + t \rightarrow m = n$, and let $m + (t + 1) = n + (t + 1)$. By the definition of addition in $\mathbb{N}$, $(m + t) + 1 = (n + t) + 1$. By the base case, $m + t = n + t$. By the inductive hypothesis, $m = n$.

By the Principle of Mathematical Induction, for all natural numbers m, n, and k, if $m + k = n + k$, then $m = n$. □

(viii)   Let $m, n \in \mathbb{N}$. We prove by induction on $k > 0$ that $mk = nk \rightarrow m = n$.

Base Case ($k = 1$): If $m \cdot 1 = n \cdot 1$, then since $m \cdot 1 = m$ and $n \cdot 1 = n$ (by part (iii) above), $m = n$.

Inductive Step: Let $t \in \mathbb{N}$, assume that $mt = nt \rightarrow m = n$, and let $m(t + 1) = n(t + 1)$. By the definition of multiplication in $\mathbb{N}$, $mt + 1 = nt + 1$. By (vii), $mt = nt$. By the inductive hypothesis, we have $m = n$.

By the Principle of Mathematical Induction, for all natural numbers m, n, and k, if $mk = nk$, then $m = n$. □

(ix)   Let $m \in \mathbb{N}$. We prove by induction on n that if $m < n$, there is $k > 0$ such that $n = m + k$.

Base Case ($t = 0$): If $m < 0$, then $m \in \emptyset$, which is impossible. So, the conclusion is vacuously true.

Inductive Step: Let $t \in \mathbb{N}$ and assume that if $m < t$, there is $k > 0$ such that $t = m + k$. Assume that $m < t + 1$. Then $m < t$ or $m = t$. If $m < t$, then $t + 1 = (m + k) + 1 = m + (k + 1)$. If $m = t$, then $t + 1 = m + 1$.

By the Principle of Mathematical Induction, for all $n \in \mathbb{N}$, if $m < n$, there is $k > 0$ so that $n = m + k$.

Now, let $n, m \in \mathbb{N}$. We prove by induction on $k > 0$ that if $n = m + k$, then $m < n$.

Base Case $(t = 1)$: If $n = m + 1 = m \cup \{m\}$, then since $m \in \{m\}$, $m \in n$, and so, $m < n$.

Inductive Step: Assume that if $n = m + t$, then $m < n$. Let $n = m + (t + 1)$. Then since addition is commutative and associative in $\mathbb{N}$, $n = m + (1 + t) = (m + 1) + t$. By the inductive hypothesis, we have $m + 1 < n$. Since $m < m + 1$ and $<$ is transitive on $\mathbb{N}$, $m < n$.

By the Principle of Mathematical Induction, if $k > 0$ and $n = m + k$, then $m < n$. $\qquad\square$

    (x)   Let $m, n, k \in \mathbb{N}$.

By (ix), $m < n$ if and only if there is a natural number $t > 0$ such that $n = m + t$. Now, $n + k = (m + t) + k = m + (t + k) = m + (k + t) = (m + k) + t$. So, if $m < n$, then there is a natural number $t > 0$ such that $n + k = (m + k) + t$. Thus, by (ix), $m + k < n + k$. Conversely, if $m + k < n + k$, then there is a natural number t such that $n + k = (m + k) + t = (m + t) + k$. By (vii), $n = m + t$. So, by (ix) again, $m < n$. $\qquad\square$

    (xi)   Let $m \in \mathbb{N}$ with $m > 0$. We prove by induction on n that $n > 0 \rightarrow mn > 0$.

Base Case $(k = 1)$: If $k = 1$, then $m \cdot 1 = m \cdot 0 + m = 0 + m = m > 0$.

Inductive Step: Let $k \in \mathbb{N}$ with $k > 0$ and assume that $mk > 0$. Then $m(k + 1) = mk + k$. Since $mk > 0$, by (x), $mk + k > 0 + k = k > 0$. So, $m(k + 1) > 0$.

By the Principle of Mathematical Induction, for all natural numbers m and n, if $m > 0$ and $n > 0$, then $mn > 0$. $\qquad\square$

5. A set A is **transitive** if $\forall x(x \in A \rightarrow x \subseteq A)$ (in words, every element of A is also a subset of A). Prove that every natural number is transitive.

Proof by Mathematical Induction:

Base Case $(k = 0)$: $0 = \emptyset$. Since $\emptyset$ has no elements, it is vacuously true that every element of $\emptyset$ is a subset of $\emptyset$.

Inductive Step: Assuming that k is transitive, let $j \in k + 1 = k \cup \{k\}$ and $m \in j$. Then $j \in k$ or $j \in \{k\}$. If $j \in k$, then we have $m \in j \in k$. Since k is transitive, $m \in k$. Therefore, $m \in k \cup \{k\} = k + 1$. If $j \in \{k\}$, then $j = k$. So, $m \in k$, and again, $m \in k \cup \{k\} = k + 1$.

By the Principle of Mathematical Induction, every natural number is transitive. $\qquad\square$

6. Determine if each of the following sequences are Cauchy sequences. Are any of the Cauchy sequences equivalent?

   (i) $(x_n) = \left(1 + \dfrac{1}{n+1}\right)$

   (ii) $(y_n) = (2^n)$

   (iii) $(z_n) = \left(1 - \dfrac{1}{2n+1}\right)$

Solutions:

   (i)   **Cauchy**

   (ii) **Not Cauchy**

   (iii) **Cauchy**

(x_n) and (z_n) are equivalent.

LEVEL 3

7. Prove that if $n \in \mathbb{N}$ and A is a nonempty subset of n, then A has a least element.

Proof by Mathematical Induction:

Base Case $(k = 0)$: $0 = \emptyset$. The only subset of $\emptyset$ is $\emptyset$. So, the statement is vacuously true.

Inductive Step: Assume that every nonempty subset of the natural number k has a least element. We will show that every nonempty subset of $k + 1 = k \cup \{k\}$ has a least element.

Let A be a nonempty subset of $k + 1$. Then $A \setminus \{k\} \subseteq k$. If $A \setminus \{k\} \neq \emptyset$, then by the inductive hypothesis, $A \setminus \{k\}$ has a least element, say j. Since $j \in k$, j is the least element of A. If $A \setminus \{k\} = \emptyset$, then k is the only element of A, and therefore, it is the least element of A.

By the Principle of Mathematical Induction, if $n \in \mathbb{N}$ and A is a nonempty subset of n, then A has a least element. $\square$

8. Prove POMI $\rightarrow$ WOP.

Proof: Assume POMI, let A be a nonempty subset of $\mathbb{N}$, and choose $n \in A$. If $n \cap A = \emptyset$, then n is the least element of A (If $m \in A$ with $m \in n$, then $m \in n \cap A$, contradicting $n \cap A = \emptyset$). Otherwise, $n \cap A$ is a nonempty subset of n, and so, by Problem 7, $n \cap A$ has a least element m. Then m is the least element of A (If $k \in A$ with $k \in m$, then $k \in n$ by Problem 5, and so, m is not the least element of $n \cap A$). $\square$

9. Prove that $<_{\mathbb{Z}}$ is a well-defined strict linear ordering on $\mathbb{Z}$. You may use the fact that $<_{\mathbb{N}}$ is a well-defined strict linear ordering on $\mathbb{N}$.

Proof: We first show that $<_{\mathbb{Z}}$ is well-defined. Suppose that $(a, b) \sim (a', b')$ and $(c, d) \sim (c', d)$. Since $(a, b) \sim (a', b')$, $a + b' = b + a'$. Since $(c, d) \sim (c', d')$, $c + d' = d + c'$.

We need to check that $[(a, b)] <_{\mathbb{Z}} [(c, d)]$ if and only if $[(a', b')] <_{\mathbb{Z}} [(c', d')]$. We have

$[(a, b)] <_{\mathbb{Z}} [(c, d)]$ if and only if $a + d <_{\mathbb{N}} b + c$ if and only if $a + d + b' + c' <_{\mathbb{N}} b + c + b' + c'$ if and only if $a + b' + d + c' <_{\mathbb{N}} b + c + b' + c'$ if and only if $b + a' + c + d' <_{\mathbb{N}} b + c + b' + c'$ if and only if $b + c + a' + d' <_{\mathbb{N}} b + c + b' + c'$ if and only if $a' + d' <_{\mathbb{N}} b' + c'$ if and only if $[(a', b')] <_{\mathbb{Z}} [(c', d')]$, as desired.

Next, we show that $<_{\mathbb{Z}}$ is antireflexive. To see this, note that $a + b \not<_{\mathbb{N}} a + b$ because $\not<_{\mathbb{N}}$ is antireflexive. So, $a + b \not<_{\mathbb{N}} b + a$. Therefore, $[(a, b)] \not<_{\mathbb{Z}} [(a, b)]$.

To see that $<_{\mathbb{Z}}$ is antisymmetric, suppose that $[(a, b)] <_{\mathbb{Z}} [(c, d)]$ and $[(c, d)] <_{\mathbb{Z}} [(a, b)]$. Then we have $a + d <_{\mathbb{N}} b + c$ and $c + b <_{\mathbb{N}} d + a$, or equivalently, $b + c <_{\mathbb{N}} a + d$. This is impossible, and so, it is vacuously true that $<_{\mathbb{Z}}$ is antisymmetric.

To see that $<_{\mathbb{Z}}$ is transitive, suppose that $[(a, b)] <_{\mathbb{Z}} [(c, d)]$ and $[(c, d)] <_{\mathbb{Z}} [(e, f)]$. Then we have $a + d <_{\mathbb{N}} b + c$ and $c + f <_{\mathbb{N}} d + e$. By adding each side of these two inequalities we get the inequality $a + d + c + f <_{\mathbb{N}} b + c + d + e$. Cancelling c and d from each side of this last inequality yields $a + f <_{\mathbb{N}} b + e$. Therefore, $[(a, b)] <_{\mathbb{Z}} [(e, f)]$.

Finally, we check that trichotomy holds. Suppose $[(a, b)] \not<_{\mathbb{Z}} [(c, d)]$ and $[(a, b)] \neq [(c, d)]$. Then $a + d \not<_{\mathbb{N}} b + c$ and $a + d \neq b + c$. Since trichotomy holds for $<_{\mathbb{N}}$, we have $b + c <_{\mathbb{N}} a + d$, or equivalently, $c + b <_{\mathbb{N}} d + a$. Therefore, $[(c, d)] <_{\mathbb{Z}} [(a, b)]$. □

LEVEL 4

10. Prove that $3^n - 1$ is even for all natural numbers n.

Proof by Mathematical Induction:

Base Case $(k = 0)$: $3^0 - 1 = 1 - 1 = 0 = 2 \cdot 0$. So, $3^0 - 1$ is even.

Inductive Step: Let $k \in \mathbb{N}$ and assume that $3^k - 1$ is even. Then $3^k - 1 = 2b$ for some integer b. Now,

$$3^{k+1} - 1 = 3^k \cdot 3^1 - 1 = 3^k \cdot 3 - 1 = 3^k \cdot 3 - 3^k + 3^k - 1 = 3^k(3 - 1) + (3^k - 1)$$

$$= 3^k \cdot 2 + 2b = 2 \cdot 3^k + 2b = 2(3^k + b).$$

Since $\mathbb{N}$ is closed under multiplication, $3^k \in \mathbb{N}$. Since $\mathbb{N}$ is closed under addition, $3^k + b \in \mathbb{N}$. Therefore, $3^{k+1} - 1$ is even.

By the Principle of Mathematical Induction, $3^n - 1$ is even for all $n \in \mathbb{N}$. □

Notes: Notice our use of SACT (see Note 7 following the proof of Theorem 4.11) in the middle of the first line of the sequence of equations. We needed $3^k - 1$ to appear, so we added 3^k, and then subtracted 3^k to the left of it.

11. Show that the Principle of Mathematical Induction is equivalent to the following statement:

(★) Let $P(n)$ be a statement and suppose that (i) $P(0)$ is true and (ii) for all $k \in \mathbb{N}$,

44

$P(k) \rightarrow P(k+1)$. Then $P(n)$ is true for all $n \in \mathbb{N}$.

Proof: Recall that the Principle of Mathematical Induction says the following: Let S be a set of natural numbers such that (i) $0 \in S$ and (ii) for all $k \in \mathbb{N}$, $k \in S \rightarrow k+1 \in S$. Then $S = \mathbb{N}$.

Suppose that the Principle of Mathematical Induction is true and let $P(n)$ be a statement such that $P(0)$ is true, and for all $k \in \mathbb{N}$, $P(k) \rightarrow P(k+1)$. Define $S = \{n \mid (P(n)\}$. Since $P(0)$ is true, $0 \in S$. If $k \in S$, then $P(k)$ is true. So, $P(k+1)$ is true, and therefore, $k + 1 \in S$. By the Principle of Mathematical Induction, $S = \mathbb{N}$. So, $P(n)$ is true for all $n \in \mathbb{N}$.

Now, suppose that $(\star)$ holds, and let S be a set of natural numbers such that $0 \in S$, and for all $k \in \mathbb{N}$, $k \in S \rightarrow k+1 \in S$. Let $P(n)$ be the statement $n \in S$. Since $0 \in S$, $P(0)$ is true. If $P(k)$ is true, then $k \in S$. So, $k + 1 \in S$, and therefore, $P(k+1)$ is true. By $(\star)$, $P(n)$ is true for all n. So, for all $n \in \mathbb{N}$, we have $n \in S$. In other words, $\mathbb{N} \subseteq S$. Since we were given $S \subseteq \mathbb{N}$, we have $S = \mathbb{N}$. $\qquad \square$

12. Prove that addition of integers is well-defined.

Proof: Suppose that $(a, b) \sim (a', b')$ and $(c, d) \sim (c', d)$. Since $(a, b) \sim (a', b')$, $a + b' = b + a'$. Since $(c, d) \sim (c', d')$, $c + d' = d + c'$.

We need to check that $(a + c, b + d) \sim (a' + c', b' + d')$, or equivalently, we need to check that $(a + c) + (b' + d') = (b + d) + (a' + c')$.

Since $a + b' = b + a'$ and $c + d' = d + c'$, we have

$$(a + c) + (b' + d') = (a + b') + (c + d') = (b + a') + (d + c') = (b + d) + (a' + c').$$

Therefore, $(a + c, b + d) = (a' + c', b' + d')$, as desired.

13. Prove that addition and multiplication of rational numbers are well-defined.

Proof: Suppose that $\frac{a}{b} = \frac{a'}{b'}$ and $\frac{c}{d} = \frac{c'}{d'}$. Since $\frac{a}{b} = \frac{a'}{b'}$, we have $ab' = ba'$. Since $\frac{c}{d} = \frac{c'}{d'}$, we have $cd' = dc'$.

We first need to check that $\frac{a}{b} + \frac{c}{d} = \frac{a'}{b'} + \frac{c'}{d'}$, or equivalently, $\frac{ad+bc}{bd} = \frac{a'd'+b'c'}{b'd'}$.

Since $ab' = ba'$ and $cd' = dc'$, we have

$$(ad + bc)(b'd') = adb'd' + bcb'd' = ab'dd' + cd'bb' = ba'dd' + dc'bb'$$
$$= bda'd' + bdb'c' = (bd)(a'd' + b'c').$$

Therefore, $\frac{ad+bc}{bd} = \frac{a'd'+b'c'}{b'd'}$, as desired.

We next need to check that $\frac{a}{b} \cdot \frac{c}{d} = \frac{a'}{b'} \cdot \frac{c'}{d'}$, or equivalently, $\frac{ac}{bd} = \frac{a'c'}{b'd'}$.

Since $ab' = ba'$ and $cd' = dc'$, we have

$$(ac)(b'd') = (ab')(cd') = (ba')(dc') = (bd)(a'c')$$

Therefore, $\frac{ac}{bd} = \frac{a'c'}{b'd''}$, as desired. □

14. Let $A = \{(x_n) \mid (x_n)$ is a Cauchy sequence of rational numbers$\}$ and define the relation R on A by $(x_n)R(y_n)$ if and only if for every $k \in \mathbb{N}^+$, there is $K \in \mathbb{N}$ such that $n > K$ implies $|x_n - y_n| < \frac{1}{k}$. Prove that R is an equivalence relation on A.

Proof: Let $(x_n) \in A$, let $k \in \mathbb{N}^+$ and let $K = 0$. Then $n > K$ implies $|x_n - x_n| = 0 < \frac{1}{k}$. So, $(x_n)R(x_n)$, and therefore, R is reflexive.

Since $|x_n - y_n| = |y_n - x_n|$, it is clear that R is symmetric.

Let $(x_n), (y_n), (z_n) \in A$ with $(x_n)R(y_n)$ and $(y_n)R(z_n)$ and let $k \in \mathbb{N}^+$. Since $(x_n)R(y_n)$, there is $K_1 \in \mathbb{N}$ such that $n > K_1$ implies $|x_n - y_n| < \frac{1}{2k}$. Since $(y_n)R(z_n)$, there is $K_2 \in \mathbb{N}$ such that $n > K_2$ implies $|y_n - z_n| < \frac{1}{2k}$. Let $K = \max\{K_1, K_2\}$. Let $n > K$. Since $K \geq K_1$, $n > K_1$, and therefore, we have $|x_n - y_n| < \frac{1}{2k}$. Since $K \geq K_2$, $n > K_2$, and therefore, we have $|y_n - z_n| < \frac{1}{2k}$. It follows that

$$|x_n - z_n| = |x_n - y_n + y_n - z_n| \leq |x_n - y_n| + |y_n - z_n| < \frac{1}{2k} + \frac{1}{2k} = 2 \cdot \frac{1}{2k} = \frac{1}{k}.$$

So, $(x_n)R(z_n)$, and therefore, R is transitive.

Since R is reflexive, symmetric, and transitive, it follows that R is an equivalence relation. □

LEVEL 5

15. The Principle of Strong Induction is the following statement:

(⋆⋆) Let $P(n)$ be a statement and suppose that (i) $P(0)$ is true and (ii) for all $k \in \mathbb{N}$, $\forall j \leq k \left(P(j)\right) \to P(k+1)$. Then $P(n)$ is true for all $n \in \mathbb{N}$.

Use the Principle of Mathematical Induction to prove the Principle of Strong Induction.

Proof: Let $P(n)$ be a statement such that $P(0)$ is true, and for all $k \in \mathbb{N}$, $\forall j \leq k \left(P(j)\right) \to P(k+1)$. Let $Q(n)$ be the statement $\forall j \leq n \left(P(j)\right)$.

Base case: $Q(0) \equiv \forall j \leq 0 (P(j)) \equiv P(0)$. Since $P(0)$ is true and $Q(0) \equiv P(0)$, $Q(0)$ is also true.

Inductive step: Suppose that $Q(k)$ is true. Then $\forall j \leq k \left(P(j)\right)$ is true. Therefore, $P(k+1)$ is true. So $Q(k) \wedge P(k+1)$ is true. But notice that

$$Q(k+1) \equiv \forall j \leq k+1 (P(j)) \equiv \forall j \leq k (P(j)) \wedge P(k+1) \equiv Q(k) \wedge P(k+1).$$

So, $Q(k+1)$ is true.

46

By the Principle of Mathematical Induction (($\star$) from Problem 11), $Q(n)$ is true for all $n \in \mathbb{N}$. This implies that $P(n)$ is true for all $n \in \mathbb{N}$. $\quad\square$

16. Use the Principle of Mathematical Induction to prove that for every $n \in \mathbb{N}$, if S is a set with $|S| = n$, then S has 2^n subsets. (Hint: Use Problem 25 from Problem Set 1.)

Proof: Base Case ($k = 0$): Let S be a set with $|S| = 0$. Then $S = \emptyset$, and the empty set has exactly 1 subset, namely itself. So, the number of subsets of S is $1 = 2^0$.

Inductive Step: Assume that for any set S with $|S| = k$, S has 2^k subsets.

Now, let A be a set with $|A| = k + 1$, let d be any element from A, and let $S = A \setminus \{d\}$ (S is the set consisting of all elements of A except d). $|S| = k$, and so, by the inductive hypothesis, S has 2^k subsets. Let $B = \{X \mid X \subseteq A \wedge d \notin X\}$ and $C = \{X \mid X \subseteq A \wedge d \in X\}$. B is precisely the set of subsets of S, and so $|B| = 2^k$. By Problem 25 from Problem Set 1, $|B| = |C|$ and therefore, $|C| = 2^k$. Also, B and C have no elements in common and every subset of A is in either B or C. So, the number of subsets of A is equal to $|B| + |C| = 2^k + 2^k = 2 \cdot 2^k = 2^1 \cdot 2^k = 2^{1+k} = 2^{k+1}$.

By the Principle of Mathematical Induction, given any $n \in \mathbb{N}$, if S is a set with $|S| = n$, then S has 2^n subsets. $\quad\square$

Notes: (1) Recall from Lesson 1 that $|S| = n$ means that the set S has n elements.

(2) Recall also from Lesson 1 that if S is a set, then the **power set** of S is the set of subsets of S.

$$\mathcal{P}(S) = \{X \mid X \subseteq S\}$$

In this problem, we proved that a set with n elements has a power set with 2^n elements. Symbolically, we have

$$|S| = n \rightarrow |\mathcal{P}(S)| = 2^n.$$

17. Prove that addition of real numbers is well-defined and that the sum of two real numbers is a real number.

Proof: Suppose that $[(x_n)] = [(z_n)]$ and $[(y_n)] = [(w_n)]$. To prove that addition is well-defined, we need to show that $[(x_n + y_n)] = [(z_n + w_n)]$.

Let $k \in \mathbb{N}^+$. Since $[(x_n)] = [(z_n)]$, there is $K_1 \in \mathbb{N}$ such that $n > K_1$ implies $|x_n - z_n| < \frac{1}{2k}$. Since $[(y_n)] = [(w_n)]$, there is $K_2 \in \mathbb{N}$ such that $n > K_2$ implies $|y_n - w_n| < \frac{1}{2k}$. Let $K = \max\{K_1, K_2\}$. Let $n > K$. Since $K \geq K_1$, $n > K_1$, and therefore, we have $|x_n - z_n| < \frac{1}{2k}$. Since $K \geq K_2$, $n > K_2$, and therefore, we have $|y_n - w_n| < \frac{1}{2k}$. It follows that

$$|(x_n + y_n) - (z_n + w_n)| = |(x_n - z_n) + (y_n - w_n)| \leq |x_n - z_n| + |y_n - w_n| < \frac{1}{2k} + \frac{1}{2k} = \frac{1}{k}.$$

So, $[(x_n + y_n)] = [(z_n + w_n)]$, as desired.

We now prove that the sum of two real numbers is a real number. Let $[(x_n)]$ and $[(y_n)]$ be real numbers. Since the sum of two rational numbers is a rational number, for each $n \in \mathbb{N}$, we have $x_n + y_n \in \mathbb{Q}$. We need to show that $(x_n + y_n)$ is a Cauchy sequence. To see this, let $k \in \mathbb{N}^+$. Since (x_n) is a Cauchy sequence, there is $K_1 \in \mathbb{N}$ such that $m \geq n > K_1$ implies $|x_m - x_n| < \frac{1}{2k}$. Since (y_n) is a Cauchy sequence, there is $K_2 \in \mathbb{N}$ such that $m \geq n > K_2$ implies $|y_m - y_n| < \frac{1}{2k}$. Let $K = \max\{K_1, K_2\}$. Suppose that $m \geq n > K$. Since $K \geq K_1$, we have $|x_m - x_n| < \frac{1}{2k}$. Since $K \geq K_2$, we have $|y_m - y_n| < \frac{1}{2k}$. So,

$$|(x_m + y_m) - (x_n + y_n)| = |(x_m - x_n) + (y_m - y_n)| \leq |x_m - x_n| + |y_m - y_n| < \frac{1}{2k} + \frac{1}{2k} = \frac{1}{k}.$$

Therefore, $(x_n + y_n)$ is a Cauchy sequence. $\qquad\square$

Problem Set 5

LEVEL 1

1. Let A and B be sets such that $A \subseteq B$. Prove that $\mathcal{P}(A) \preccurlyeq \mathcal{P}(B)$.

Proof: Suppose that $A \subseteq B$. We show that $\mathcal{P}(A) \subseteq \mathcal{P}(B)$. Let $X \in \mathcal{P}(A)$. Then $X \subseteq A$. Since $X \subseteq A$ and $A \subseteq B$, and $\subseteq$ is transitive (Theorem 1.14), we have $X \subseteq B$. Therefore, $X \in \mathcal{P}(B)$. Since X was an arbitrary element of $\mathcal{P}(A)$, we have shown that every element of $\mathcal{P}(A)$ is an element of $\mathcal{P}(B)$. Therefore, $\mathcal{P}(A) \subseteq \mathcal{P}(B)$. By Note 1 following Example 5.6, $\mathcal{P}(A) \preccurlyeq \mathcal{P}(B)$. $\qquad\square$

2. Let A, B, and C be sets. Prove the following:

    (i) $\preccurlyeq$ is transitive.

    (ii) $\prec$ is transitive.

    (iii) If $A \preccurlyeq B$ and $B \prec C$, then $A \prec C$.

    (iv) If $A \prec B$ and $B \preccurlyeq C$, then $A \prec C$.

Proofs:

    (i)   Suppose that $A \preccurlyeq B$ and $B \preccurlyeq C$. Then there are functions $f : A \hookrightarrow B$ and $g : B \hookrightarrow C$. By Theorem 3.5, $g \circ f : A \hookrightarrow C$. So, $A \preccurlyeq C$. Therefore, $\preccurlyeq$ is transitive. $\qquad\square$

    (ii)   Suppose that $A \prec B$ and $B \prec C$. Then $A \preccurlyeq B$ and $B \preccurlyeq C$. By (i), $A \preccurlyeq C$. Assume toward contradiction that $A \sim C$. Since $\sim$ is symmetric, $C \sim A$. In particular, $C \preccurlyeq A$. Since $C \preccurlyeq A$ and $A \preccurlyeq B$, by (i), $C \preccurlyeq B$. Since $B \preccurlyeq C$ and $C \preccurlyeq B$, by the Cantor-Schroeder-Bernstein Theorem, $B \sim C$, contradicting $B \prec C$. It follows that $A \nsim C$, and thus, $A \prec C$. $\qquad\square$

    (iii) Suppose that $A \preccurlyeq B$ and $B \prec C$. Then $B \preccurlyeq C$. By (i), $A \preccurlyeq C$. Assume toward contradiction that $A \sim C$. The rest of the argument is the same as (ii). $\qquad\square$

    (iv) Suppose that $A \prec B$ and $B \preccurlyeq C$. Then $A \preccurlyeq B$. By (i), $A \preccurlyeq C$. Assume toward contradiction that $A \sim C$. Since $\sim$ is symmetric, $C \sim A$. In particular, $C \preccurlyeq A$. Since $B \preccurlyeq C$ and $C \preccurlyeq A$, by (i), $B \preccurlyeq A$. Since $A \preccurlyeq B$ and $B \preccurlyeq A$, by the Cantor-Schroeder-Bernstein Theorem, $A \sim B$, contradicting $A \prec B$. It follows that $A \nsim C$, and thus, $A \prec C$. $\qquad\square$

LEVEL 2

3. Define $\mathcal{P}_k(\mathbb{N})$ for each $k \in \mathbb{N}$ by $\mathcal{P}_0(\mathbb{N}) = \mathbb{N}$ and $\mathcal{P}_{k+1}(\mathbb{N}) = \mathcal{P}\big(\mathcal{P}_k(\mathbb{N})\big)$ for $k > 0$. Find a set B such that for all $k \in \mathbb{N}$, $\mathcal{P}_k(\mathbb{N}) \prec B$.

Solution: Let $B = \bigcup \{\mathcal{P}_n(\mathbb{N}) \mid n \in \mathbb{N}\}$. Let $k \in \mathbb{N}$. Since $\mathcal{P}_k(\mathbb{N}) \subseteq B$, by Note 1 following Example 5.6, $\mathcal{P}_k(\mathbb{N}) \preccurlyeq B$. Since k was arbitrary, we have $\mathcal{P}_k(\mathbb{N}) \preccurlyeq B$ for all $k \in \mathbb{N}$. Again, let $k \in \mathbb{N}$. We have $\mathcal{P}_k(\mathbb{N}) \prec \mathcal{P}_{k+1}(\mathbb{N})$ and $\mathcal{P}_{k+1}(\mathbb{N}) \preccurlyeq B$. By Problem 2 (part (iv)), $\mathcal{P}_k(\mathbb{N}) \prec B$. Since $k \in \mathbb{N}$ was arbitrary, we have shown that for all $k \in \mathbb{N}$, $\mathcal{P}_k(\mathbb{N}) \prec B$.

4. Prove that if $A \sim B$ and $C \sim D$, then $A \times C \sim B \times D$.

49

Proof: Suppose that $A \sim B$ and $C \sim D$. Then there exist bijections $h: A \to B$ and $k: C \to D$. Define $f: A \times C \to B \times D$ by $f(a,c) = \big(h(a), k(c)\big)$.

Suppose $(a,c), (a',c') \in A \times C$ with $f\big((a,c)\big) = f\big((a',c')\big)$. Then $\big(h(a), k(c)\big) = \big(h(a'), k(c')\big)$. So, $h(a) = h(a')$ and $k(c) = k(c')$. Since h is an injection, $a = a'$. Since k is an injection, $c = c'$. Since $a = a'$ and $c = c'$, $(a,c) = (a',c')$. Since $(a,c), (a',c') \in A \times C$ were arbitrary, f is an injection.

Now, let $(b,d) \in B \times D$. Since h and k are bijections, h^{-1} and k^{-1} exist. Let $a = h^{-1}(b)$, $c = k^{-1}(d)$. Then $f(a,c) = \big(h(a), k(c)\big) = \Big(h\big(h^{-1}(b)\big), k\big(k^{-1}(d)\big)\Big) = (b,d)$. Since $(b,d) \in B \times D$ was arbitrary, f is a surjection.

Since f is both an injection and a surjection, $A \times C \sim B \times D$. $\qquad\square$

LEVEL 3

5. Prove the following:

   (i)   There is a partition P of $\mathbb{N}$ such that $P \sim \mathbb{N}$ and for each $X \in P$, $X \sim \mathbb{N}$

   (ii) A countable union of countable sets is countable.

Proofs:

(i)   For each $n \in \mathbb{N}$, let P_n be the set of natural numbers ending with exactly n zeros and let $P = \{P_n \mid n \in \mathbb{N}\}$. For example, $5231 \in P_0$, $0 \in P_1$, and $26{,}200 \in P_2$. Let's define $\widetilde{m,n}$ to be the natural number consisting of m 1's followed by n 0's. For example, $\widetilde{3,0} = 111$ and $\widetilde{2,5} = 1{,}100{,}000$. For each $n \in \mathbb{N}$, $\{\widetilde{m,n} \mid m \in \mathbb{N}\} \subseteq P_n$ showing that each P_n is equinumerous to $\mathbb{N}$. Also, if $k \in P_n \cap P_m$, then k ends with exactly n zeros and exactly m zeros, and so, $n = m$. Therefore, P is pairwise disjoint. This also shows that the function $f: \mathbb{N} \to P$ defined by $f(n) = P_n$ is a bijection. So, $P \sim \mathbb{N}$. Finally, if $k \in \mathbb{N}$, then there is $n \in \mathbb{N}$ such that k ends with exactly n zeros. So, $\bigcup P = \mathbb{N}$. $\qquad\square$

(ii) For each $n \in \mathbb{N}$, let A_n be a countable set. By replacing each A_n by $A_n \times \{n\}$, we can assume that $\{A_n \mid n \in \mathbb{N}\}$ is a pairwise disjoint collection of sets ($A_n \sim A_n \times \{n\}$ via the bijection f sending x to (x, n)). By (i), there is a partition P of $\mathbb{N}$ such that $P \sim \mathbb{N}$ and for each $X \in P$, $X \sim \mathbb{N}$. Let's say $P = \{P_n \mid n \in \mathbb{N}\}$. Since each A_n is countable, for each $n \in \mathbb{N}$ there are injective functions $f_n: A_n \to P_n$. Define $f: \bigcup\{A_n \mid n \in \mathbb{N}\} \to \mathbb{N}$ by $f(x) = f_n(x)$ if $x \in A_n$.

Since $\{A_n \mid n \in \mathbb{N}\}$ is pairwise disjoint, f is well-defined.

Suppose that $x, y \in \bigcup\{A_n \mid n \in \mathbb{N}\}$ with $f(x) = f(y)$. There exist $n, m \in \mathbb{N}$ such that $x \in A_n$ and $y \in A_m$. So, $f(x) = f_n(x) \in P_n$ and $f(y) = f_m(y) \in P_m$. Since $f(x) = f(y)$, we have $f_n(x) = f_m(y)$. Since for $n \neq m$, $P_n \cap P_m = \emptyset$, we must have $n = m$. So, we have $f_n(x) = f_n(y)$. Since f_n is injective, $x = y$. Since $x, y \in \bigcup\{A_n \mid n \in \mathbb{N}\}$ were arbitrary, f is an injective function. Therefore, $\bigcup\{A_n \mid n \in \mathbb{N}\}$ is countable. $\qquad\square$

6. Let A and B be sets such that $A \sim B$. Prove that $\mathcal{P}(A) \sim \mathcal{P}(B)$.

Proof: Suppose that $A \sim B$. Then there exists a bijection $h: A \to B$. Define $F: \mathcal{P}(A) \to \mathcal{P}(B)$ by $F(X) = \{h(a) \mid a \in X\}$ for each $X \in \mathcal{P}(A)$.

Suppose $X, Y \in \mathcal{P}(A)$ with $F(X) = F(Y)$. Let $a \in X$. Then $h(a) \in F(X)$. Since $F(X) = F(Y)$, $h(a) \in F(Y)$. So, there is $b \in Y$ such that $h(a) = h(b)$. Since h is injective, $a = b$. So, $a \in Y$. Since $a \in X$ was arbitrary, $X \subseteq Y$. By a symmetrical argument, $Y \subseteq X$. Therefore, $X = Y$. Since $X, Y \in \mathcal{P}(A)$ were arbitrary, F is injective.

Let $Y \in \mathcal{P}(B)$ and let $X = \{a \in A \mid h(a) \in Y\}$. Then $b \in F(X)$ if and only if $b = h(a)$ for some $a \in X$ if and only if $b \in Y$ (because h is surjective). So, $F(X) = Y$. Since $Y \in \mathcal{P}(B)$ was arbitrary, F is surjective.

Since F is injective and surjective, $\mathcal{P}(A) \sim \mathcal{P}(B)$. $\qquad\square$

LEVEL 4

7. Prove the following:

   (i) $\mathbb{N} \times \mathbb{N} \sim \mathbb{N}$.

   (ii) $\mathbb{Q} \sim \mathbb{N}$.

   (iii) Any two intervals of real numbers are equinumerous (including $\mathbb{R}$ itself).

   (iv) ${}^{\mathbb{N}}\mathbb{N} \sim \mathcal{P}(\mathbb{N})$.

Proofs:

(i) $\mathbb{N} \times \mathbb{N} = \bigcup\{\mathbb{N} \times \{n\} \mid n \in \mathbb{N}\}$. This is a countable union of countable sets. By Problem 5 (part(ii)), $\mathbb{N} \times \mathbb{N}$ is countable. $\qquad\square$

(ii) $\mathbb{Q}^+ = \left\{\frac{a}{b} \,\middle|\, a \in \mathbb{N} \wedge b \in \mathbb{N}^+\right\} = \bigcup\left\{\left\{\frac{a}{b} \,\middle|\, a \in \mathbb{N}\right\} \,\middle|\, b \in \mathbb{N}^+\right\}$. This is a countable union of countable sets. By Problem 5 (part(ii)), $\mathbb{Q}^+$ is countable. Now, $\mathbb{Q} = \mathbb{Q}^+ \cup \{0\} \cup \mathbb{Q}^-$, where $\mathbb{Q}^- = \{q \in \mathbb{Q} \mid -q \in \mathbb{Q}^+\}$. This is again a countable union of countable sets, thus countable. So, $\mathbb{Q} \sim \mathbb{N}$. $\qquad\square$

(iii) The function $f: \mathbb{R} \to (0, \infty)$ defined by $f(x) = 2^x$ is a bijection. So, $\mathbb{R} \sim (0, \infty)$. The function $g: (0, \infty) \to (0, 1)$ defined by $g(x) = \frac{1}{x^2+1}$ is a bijection. So, $(0, \infty) \sim (0, 1)$. If $a, b \in \mathbb{R}$, the function $h: (0, 1) \to (a, b)$ defined by $h(x) = (b - a)x + a$ is a bijection. So, $(0, 1) \sim (a, b)$. It follows that all bounded open intervals are equinumerous with each other and $\mathbb{R}$.

We have, $[a, b] \subseteq (a - 1, b + 1) \sim (a, b) \subseteq [a, b) \subseteq [a, b]$ and $(a, b) \subseteq (a, b] \subseteq [a, b]$. It follows that all bounded intervals are equinumerous with each other and $\mathbb{R}$.

We also have the following.

$$(a, \infty) \subseteq [a, \infty) \subseteq \mathbb{R} \sim (a, a + 1) \subseteq (a, \infty)$$

$$(-\infty, b) \subseteq (-\infty, b] \subseteq \mathbb{R} \sim (b - 1, b) \subseteq (-\infty, b)$$

Therefore, all unbounded intervals are equinumerous with $\mathbb{R}$. It follows that any two intervals of real numbers are equinumerous. $\qquad\square$

(iv) $^{\mathbb{N}}\mathbb{N} \subseteq \mathcal{P}(\mathbb{N} \times \mathbb{N})$ by the definition of $^{\mathbb{N}}\mathbb{N}$. So, $^{\mathbb{N}}\mathbb{N} \preccurlyeq \mathcal{P}(\mathbb{N} \times \mathbb{N})$ by Note 1 following Example 5.6. By (i) above, $\mathbb{N} \times \mathbb{N} \sim \mathbb{N}$. So, by Problem 6, $\mathcal{P}(\mathbb{N} \times \mathbb{N}) \sim \mathcal{P}(\mathbb{N})$. Thus, $\mathcal{P}(\mathbb{N} \times \mathbb{N}) \preccurlyeq \mathcal{P}(\mathbb{N})$. Since $\preccurlyeq$ is transitive, $^{\mathbb{N}}\mathbb{N} \preccurlyeq \mathcal{P}(\mathbb{N})$.

Now, $\mathcal{P}(\mathbb{N}) \sim {}^{\mathbb{N}}\{0, 1\}$ (see Example 5.1 (part 5)). So, $\mathcal{P}(\mathbb{N}) \preccurlyeq {}^{\mathbb{N}}\{0, 1\}$. Also, $^{\mathbb{N}}\{0, 1\} \subseteq {}^{\mathbb{N}}\mathbb{N}$, and so, by Note 1 following Example 5.6, $^{\mathbb{N}}\{0, 1\} \preccurlyeq {}^{\mathbb{N}}\mathbb{N}$. Since $\preccurlyeq$ is transitive, $\mathcal{P}(\mathbb{N}) \preccurlyeq {}^{\mathbb{N}}\mathbb{N}$.

By the Cantor-Schroeder-Bernstein Theorem, $^{\mathbb{N}}\mathbb{N} \sim \mathcal{P}(\mathbb{N})$.   □

Notes: (1) In the proof of (iii), we used the fact that equinumerosity is an equivalence relation, the Cantor-Schroeder-Bernstein Theorem, and Problem 2 many times without mention. For example, we have $\mathbb{R} \sim (0, \infty)$ and $(0, \infty) \sim (0, 1)$. So, by the transitivity of $\sim$, we have $\mathbb{R} \sim (0, 1)$. As another example, the sequence $(a, \infty) \subseteq [a, \infty) \subseteq \mathbb{R} \sim (a, a + 1) \subseteq (a, \infty)$ together with Note 1 following Example 5.6 gives us that $(a, \infty) \preccurlyeq \mathbb{R}$ and $\mathbb{R} \preccurlyeq (a, \infty)$. By the Cantor-Schroeder-Bernstein Theorem, $(a, \infty) \sim \mathbb{R}$.

(2) Once we showed that for all $a, b \in \mathbb{R}$, $(0, 1) \sim (a, b)$, it follows from the fact that $\sim$ is an equivalence relation that any two bounded open intervals are equinumerous. Indeed, if (a, b) and (c, d) are bounded open intervals, then $(0, 1) \sim (a, b)$ and $(0, 1) \sim (c, d)$. By the symmetry of $\sim$, we have $(a, b) \sim (0, 1)$, and finally, by the transitivity of $\sim$, we have $(a, b) \sim (c, d)$.

(3) It's easy to prove that two specific intervals of real numbers are equinumerous using just the fact that any two bounded open intervals are equinumerous with each other, together with the fact that $\mathbb{R} \sim (0, 1)$. For example, to show that $[3, \infty)$ is equinumerous with $(-2, 5]$, simply consider the following sequence: $[3, \infty) \subseteq \mathbb{R} \sim (0, 1) \sim (-2, 5) \subseteq (-2, 5] \subseteq (-2, 6) \sim (3, 4) \subseteq [3, \infty)$.

(4) In the proof of (iii), we claimed that the function $f: \mathbb{R} \rightarrow (0, \infty)$ defined by $f(x) = 2^x$ is a bijection without proof. In fact, at this point, we do not even have a definition of 2^x for all real values of x. The formal definition and theory of 2^x will be provided in Lesson 12.

8.   Prove that if $A \sim B$ and $C \sim D$, then $^{A}C \sim {}^{B}D$.

Proof: Suppose that $A \sim B$ and $C \sim D$. Then there exist bijections $h: A \rightarrow B$ and $k: C \rightarrow D$. Define $F: {}^{A}C \rightarrow {}^{B}D$ by $F(f)(b) = k\left(f\left(h^{-1}(b)\right)\right)$.

Suppose $f, g \in {}^{A}C$ with $F(f) = F(g)$. Let $a \in A$, and let $b = h(a)$. We have $F(f)(b) = F(g)(b)$, or equivalently, $k\left(f\left(h^{-1}(b)\right)\right) = k\left(g\left(h^{-1}(b)\right)\right)$. Since k is injective, $f\left(h^{-1}(b)\right) = g\left(h^{-1}(b)\right)$. Since $b = h(a)$, $a = h^{-1}(b)$. So, $f(a) = g(a)$. Since $a \in A$ was arbitrary, $f = g$. Since $f, g \in {}^{A}C$ were arbitrary, F is injective.

Now, let $g \in {}^{B}D$ and let's define $f \in {}^{A}C$ by $f(a) = k^{-1}\left(g\left(h(a)\right)\right)$. Let $b \in B$. Then we have

$$F(f)(b) = k\left(f\left(h^{-1}(b)\right)\right) = k\left(k^{-1}\left(g\left(h\left(h^{-1}(b)\right)\right)\right)\right) = g(b).$$ Since $b \in B$ was arbitrary, we have $F(f) = g$. Since $g \in {}^{B}D$ was arbitrary, F is surjective.

Since F is injective and surjective, $^{A}C \sim {}^{B}D$.   □

LEVEL 5

9. Prove that for any sets A, B, and C, $^{B \times C}A \sim {}^{C}(^{B}A)$.

Proof: Let A, B, and C be sets, and define $F: {}^{B \times C}A \to {}^{C}(^{B}A)$ by $F(f)(c)(b) = f(b, c)$.

Suppose $f, g \in {}^{B \times C}A$ with $F(f) = F(g)$. Let $c \in C$. Since $F(f) = F(g)$, $F(f)(c) = F(g)(c)$. So, for all $b \in B$, $F(f)(c)(b) = F(g)(c)(b)$. So, for all $b \in B$, $f(b, c) = g(b, c)$. Since $c \in C$ was arbitrary, for all $b \in B$ and $c \in C$, $f(b, c) = g(b, c)$. Therefore, $f = g$. Since $f, g \in {}^{B \times C}A$ were arbitrary, F is injective.

Let $k \in {}^{C}(^{B}A)$ and define $f \in {}^{B \times C}A$ by $f(b, c) = k(c)(b)$. Then $F(f)(c)(b) = f(b, c) = k(c)(b)$. So, $F(f) = k$. Since $k \in {}^{C}(^{B}A)$ was arbitrary, F is surjective.

Since F is injective and surjective, $^{B \times C}A \sim {}^{C}(^{B}A)$. □

10. Prove the following:

    (i) $\mathcal{P}(\mathbb{N}) \sim \{f \in {}^{\mathbb{N}}\mathbb{N} \mid f \text{ is a bijection}\}$.

    (ii) $\mathbb{R} \sim \mathcal{P}(\mathbb{N})$.

    (iii) $^{\mathbb{N}}\mathbb{R} \not\sim {}^{\mathbb{R}}\mathbb{N}$.

Proofs:

(i)   Let $S = \{f \in {}^{\mathbb{N}}\mathbb{N} \mid f \text{ is a bijection}\}$. Then $S \subseteq {}^{\mathbb{N}}\mathbb{N}$. So $S \preccurlyeq {}^{\mathbb{N}}\mathbb{N}$ by Note 1 following Example 5.6. By part (iv) of Problem 7, $^{\mathbb{N}}\mathbb{N} \sim \mathcal{P}(\mathbb{N})$. So, $^{\mathbb{N}}\mathbb{N} \preccurlyeq \mathcal{P}(\mathbb{N})$. By the transitivity of $\preccurlyeq$, $S \preccurlyeq \mathcal{P}(\mathbb{N})$.

Now, define $F: \mathcal{P}(\mathbb{N}) \to S$ by $F(A) = f_A$, where f_A is defined as follows: if $n \notin A$, then $f_A(2n) = 2n$ and $f_A(2n + 1) = 2n + 1$; if $n \in A$, then $f_A(2n) = 2n + 1$ and $f_A(2n + 1) = 2n$.

To see that F is injective, suppose that $A, B \in \mathcal{P}(\mathbb{N})$ and $A \neq B$. Without loss of generality, suppose that there is $n \in A \setminus B$. Then $f_A(2n) = 2n + 1$ and $f_B(2n) = 2n$. So, $f_A \neq f_B$. Thus, $F(A) \neq F(B)$, and therefore, F is injective.

Since $S \preccurlyeq \mathcal{P}(\mathbb{N})$ and $\mathcal{P}(\mathbb{N}) \preccurlyeq S$, by the Cantor-Schroeder-Bernstein Theorem, $\mathcal{P}(\mathbb{N}) \sim S$. □

(ii)   We first show that $\mathbb{R} \preccurlyeq {}^{\mathbb{N}}\mathbb{Q}$. Define $f: \mathbb{R} \to {}^{\mathbb{N}}\mathbb{Q}$ by $f(x) = (x_n)$, where (x_n) is some representative of x. Clearly, f is injective. It follows that

$$\mathbb{R} \preccurlyeq {}^{\mathbb{N}}\mathbb{Q} \subseteq \mathcal{P}(\mathbb{N} \times \mathbb{Q}) \sim \mathcal{P}(\mathbb{N} \times \mathbb{N}) \sim \mathcal{P}(\mathbb{N}).$$

Thus, $\mathbb{R} \preccurlyeq \mathcal{P}(\mathbb{N})$.

Next, we show that $\mathcal{P}(\mathbb{N}) \preccurlyeq \mathbb{R}$. Let $A \in \mathcal{P}(\mathbb{N})$. For each $n \in \mathbb{N}$, let $a_n = 0$ if $n \notin A$ and let $a_n = 1$ if $n \in A$. Let $a_n^* = \frac{a_0 a_1 \cdots a_n}{10^n}$. Then (a_n^*) is a Cauchy sequence (Check this!). Define $g: \mathcal{P}(\mathbb{N}) \to \mathbb{R}$ by $g(A) = [(a_n^*)]$. Then g is injective (Check this!).

Since $\mathbb{R} \preccurlyeq \mathcal{P}(\mathbb{N})$ and $\mathcal{P}(\mathbb{N}) \preccurlyeq \mathbb{R}$, by the Cantor-Schroeder-Bernstein Theorem, we have $\mathbb{R} \sim \mathcal{P}(\mathbb{N})$. □

(iii)   By (ii), $\mathbb{R} \sim \mathcal{P}(\mathbb{N})$. By Problem 8, $^{\mathbb{N}}\mathbb{R} \sim {}^{\mathbb{N}}\mathcal{P}(\mathbb{N})$.

Using previous equinumerosity results, we get the following:

$$^{\mathbb{N}}\mathbb{R} \sim {}^{\mathbb{N}}\mathcal{P}(\mathbb{N}) \sim {}^{\mathbb{N}}({}^{\mathbb{N}}2) \sim {}^{\mathbb{N}\times\mathbb{N}}2 \sim {}^{\mathbb{N}}2 \sim \mathcal{P}(\mathbb{N}) \sim \mathbb{R} \prec \mathcal{P}(\mathbb{R}) \sim {}^{\mathbb{R}}2 \subseteq {}^{\mathbb{R}}\mathbb{N}.$$

It follows that $^{\mathbb{N}}\mathbb{R} \prec {}^{\mathbb{R}}\mathbb{N}$. $\square$

Note: To help us understand the function F defined in part (i) above, let's draw a visual representation of $F(\mathbb{E})$, where $\mathbb{E}$ is the set of even natural numbers.

$$\mathbb{E} \qquad F(\mathbb{E}) = f_{\mathbb{E}}$$

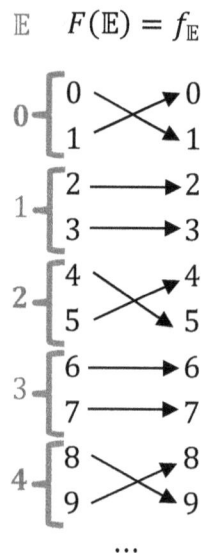

Along the left of the image we have listed the natural numbers 0, 1, 2, 3, 4, ... (we stopped at 4, but our intention is that they keep going). The elements of $\mathbb{E}$ are 0, 2, 4, ... We highlighted these in bold. We associate each natural number n with the pair $\{2n, 2n + 1\}$. For example, $2 \cdot 4 = 8$ and $2 \cdot 4 + 1 = 9$. So, we associate 4 with the pair of natural numbers $\{8, 9\}$. We used left braces to indicate that association. The arrows give a visual representation of $f_{\mathbb{E}}$. Since $0 \in \mathbb{E}$, $f_{\mathbb{E}}$ swaps the corresponding pair 0 and 1. Since $1 \notin \mathbb{E}$, $f_{\mathbb{E}}$ leaves the corresponding pair 2 and 3 fixed. And so on, down the line...

The configuration of $f_{\mathbb{O}}$, where $\mathbb{O}$ is the set of odd natural numbers would be the opposite of the configuration for the evens. For example, 0 and 1 would remain fixed, while 2 and 3 would be swapped.

Problem Set 6

LEVEL 1

1. Show that there are exactly two monoids on the set $S = \{e, a\}$, where e is the identity. Which of these monoids are groups? Which of these monoids are commutative?

Solution: Let's let e be the identity. Since $e \star x = x \star e = x$ for all x in the monoid, we can fill out the first row and the first column of the table as follows:

$\star$	e	a
e	e	a
a	a	$\boxdot$

Now, the entry labeled with $\boxdot$ must be either e or a because we need $\star$ to be a binary operation on S.

Case 1: If we let $\boxdot$ be a, we get the following table.

$\star$	e	a
e	e	a
a	a	a

Associativity holds because any computation of the form $(x \star y) \star z$ or $x \star (y \star z)$ will result in a if any of x, y, or z is a. So, all that is left to check is that $(e \star e) \star e = e \star (e \star e)$. But each side of that equation is equal to e.

So, with this multiplication table, $(S, \star)$ **is** a monoid.

This monoid is **not** a group because a has no inverse. Indeed, $a \star e = a \neq e$ and $a \star a = a \neq e$.

This monoid **is** commutative because $a \star e = a$ and $e \star a = a$.

Case 2: If we let $\boxdot$ be e, we get the following table.

$\star$	e	a
e	e	a
a	a	e

Let's check that associativity holds. There are eight instances to check.

$$(e \star e) \star e = e \star e = e \qquad e \star (e \star e) = e \star e = e$$
$$(e \star e) \star a = e \star a = a \qquad e \star (e \star a) = e \star a = a$$
$$(e \star a) \star e = a \star e = a \qquad e \star (a \star e) = e \star a = a$$
$$(a \star e) \star e = a \star e = a \qquad a \star (e \star e) = a \star e = a$$
$$(e \star a) \star a = a \star a = e \qquad e \star (a \star a) = e \star e = e$$
$$(a \star e) \star a = a \star a = e \qquad a \star (e \star a) = a \star a = e$$
$$(a \star a) \star e = e \star e = e \qquad a \star (a \star e) = a \star a = e$$
$$(a \star a) \star a = e \star a = a \qquad a \star (a \star a) = a \star e = a$$

So, with this multiplication table, $(S, \star)$ **is** a monoid.

Since $e \star e = e$, e is its own inverse. Since $a \star a = e$, a is also its own inverse. Therefore, each element of this monoid is invertible. It follows that this monoid **is** a group.

This monoid **is** commutative because $a \star e = a$ and $e \star a = a$.

2. The addition and multiplication tables below are defined on the set $S = \{0, 1\}$. Show that $(S, +, \cdot)$ does **not** define a ring.

+	0	1
0	0	1
1	1	0

$\cdot$	0	1
0	1	0
1	0	1

Solution: We have $0(1 + 1) = 0 \cdot 0 = 1$ and $0 \cdot 1 + 0 \cdot 1 = 0 + 0 = 0$. So, $0(1 + 1) \neq 0 \cdot 1 + 0 \cdot 1$. Therefore, multiplication is **not** distributive over addition in S, and so, $(S, +, \cdot)$ does not define a ring.

Notes: (1) Both multiplication tables given are the same, except that we interchanged the roles of 0 and 1 (in technical terms, $(S, +)$ and $(S, \cdot)$ are **isomorphic**).

Both tables represent the unique table for a group with 2 elements. See Problem 1 above for details.

(2) Since $(S, +)$ is a commutative group and $(S, \cdot)$ is a monoid (in fact, it's a commutative group), we know that the only possible way $(S, +, \cdot)$ can fail to be a ring is for distributivity to fail.

3. The addition and multiplication tables below are defined on the set $S = \{0, 1, 2\}$. Show that $(S, +, \cdot)$ does **not** define a field.

+	0	1	2
0	0	1	2
1	1	2	0
2	2	0	1

$\cdot$	0	1	2
0	0	0	0
1	0	1	2
2	0	2	2

Solution: We have $2 \cdot 0 = 0$, $2 \cdot 1 = 2$, and $2 \cdot 2 = 2$. So, 2 has no multiplicative inverse, and therefore, $(S, +, \cdot)$ does **not** define a field.

Note: It's not difficult to check that $(S, +)$ is a group with identity 0 and $(S, \cdot)$ is a monoid with identity 1. However, $(S, +, \cdot)$ is not a ring, as distributivity fails. Here is a counterexample:

$$2(1 + 1) = 2 \cdot 2 = 2 \qquad 2 \cdot 1 + 2 \cdot 1 = 2 + 2 = 1$$

We could have used this computation to verify that $(S, +, \cdot)$ is not a field.

4. Let $F = \{0, 1\}$, where $0 \neq 1$. Show that there is exactly one field $(F, +, \cdot)$, where 0 is the additive identity and 1 is the multiplicative identity.

Solution: Suppose that $(F, +, \cdot)$ is a field. Since $(F, +)$ is a commutative group, by Problem 1 above, the addition table must be the following.

+	0	1
0	0	1
1	1	0

Since $(F^*, \cdot)$ is a monoid and 1 is the multiplicative identity, we must have $1 \cdot 1 = 1$.

Now, if $0 \cdot 0 = 1$, then we have $1 = 0 \cdot 0 = 0(0 + 0) = 0 \cdot 0 + 0 \cdot 0 = 1 + 1 = 0$, a contradiction. So, $0 \cdot 0 = 0$.

If $0 \cdot 1 = 1$, then we have $1 = 0 \cdot 1 = (0 + 0) \cdot 1 = 0 \cdot 1 + 0 \cdot 1 = 1 + 1 = 0$, a contradiction. So, $0 \cdot 1 = 0$.

Finally, if $1 \cdot 0 = 1$, then we have $1 = 1 \cdot 0 = 1(0 + 0) = 1 \cdot 0 + 1 \cdot 0 = 1 + 1 = 0$, a contradiction. So, $1 \cdot 0 = 0$.

It follows that the addition and multiplication tables must be as follows:

+	0	1
0	0	1
1	1	0

$\cdot$	0	1
0	0	0
1	0	1

Since we already know that $(F, +)$ is a commutative group and $(F^*, \cdot)$ is a monoid, all we need to verify is that distributivity and the multiplicative inverse property hold. Since $\cdot$ is commutative for S (by Problem 1 above), it suffices to verify left distributivity. We will do this by brute force. There are eight instances to check.

$$0(0 + 0) = 0 \cdot 0 = 0 \qquad 0 \cdot 0 + 0 \cdot 0 = 0 + 0 = 0$$
$$0(0 + 1) = 0 \cdot 1 = 0 \qquad 0 \cdot 0 + 0 \cdot 1 = 0 + 0 = 0$$
$$0(1 + 0) = 0 \cdot 1 = 0 \qquad 0 \cdot 1 + 0 \cdot 0 = 0 + 0 = 0$$
$$0(1 + 1) = 0 \cdot 0 = 0 \qquad 0 \cdot 1 + 0 \cdot 1 = 0 + 0 = 0$$
$$1(0 + 0) = 1 \cdot 0 = 0 \qquad 1 \cdot 0 + 1 \cdot 0 = 0 + 0 = 0$$
$$1(0 + 1) = 1 \cdot 1 = 1 \qquad 1 \cdot 0 + 1 \cdot 1 = 0 + 1 = 1$$
$$1(1 + 0) = 1 \cdot 1 = 1 \qquad 1 \cdot 1 + 1 \cdot 0 = 1 + 0 = 1$$
$$1(1 + 1) = 1 \cdot 0 = 0 \qquad 1 \cdot 1 + 1 \cdot 1 = 1 + 1 = 0$$

So, we see that left distributivity holds, and therefore $(S, +, \cdot)$ is a commutative ring.

Since $1 \cdot 1 = 1$, the multiplicative inverse property holds, and it follows that $(F, +, \cdot)$ is a field.

LEVEL 2

5. Let $G = \{e, a, b\}$ and let $(G, \star)$ be a group with identity element e. Draw a multiplication table for $(G, \star)$.

Solution: Since $e \star x = x \star e = x$ for all x in the group, we can fill out the first row and the first column of the table as follows:

$\star$	e	a	b
e	e	a	b
a	a	$\boxdot$	
b	b		

Now, the entry labeled with ⊡ must be either e or b because a is already in that row. If it were e, then the final entry in the row would be b giving two b's in the last column. Therefore, the entry labeled with ⊡ must be b.

$\star$	e	a	b
e	e	a	b
a	a	b	
b	b		

Since the same element cannot be repeated in any row or column, the rest of the table is now determined.

$\star$	e	a	b
e	e	a	b
a	a	b	e
b	b	e	a

Notes: (1) Why can't the same element appear twice in any row? Well if x appeared twice in the row corresponding to y, that would mean that there are elements z and w with $z \neq w$ such that $y \star z = x$ and $y \star w = x$. So, $y \star z = y \star w$. We can multiply each side of the equation on the left by y^{-1} (the inverse of y) to get $y^{-1} \star (y \star z) = y^{-1} \star (y \star w)$. By associativity, $(y^{-1} \star y) \star z = (y^{-1} \star y) \star w$. Now, $y^{-1} \star y = e$ by the inverse property. So, we have $e \star z = e \star w$. Finally, since e is an identity, $z = w$. This contradiction establishes that no element x can appear twice in the same row of a group multiplication table.

A similar argument can be used to show that the same element cannot appear twice in any column.

(2) The argument given in Note 1 used all the group properties (associativity, identity, and inverse). What if we remove one of the properties. For example, what about the multiplication table for a monoid? Can an element appear twice in a row or column? See the solution to Problem 1 above for the answer to this question.

(3) In Note 1 above, we showed that in the multiplication table for a group, the same element cannot appear as the output more than once in any row or column. We can also show that every element must appear in every row and column. Let's show that the element y must appear in the row corresponding to x. We are looking for an element z such that $x \star z = y$. Well, $z = x^{-1} \star y$ works. Indeed, we have $x \star (x^{-1} \star y) = (x \star x^{-1}) \star y = e \star y = y$.

(4) Using Notes 1 and 3, we see that each element of a group appears exactly once in every row and column of the group's multiplication table.

(5) We have shown that there is essentially just one group of size 3, namely the one given by the table that we produced. Any other group with 3 elements will look exactly like this one, except for possibly the names of the elements. In technical terms, we say that any two groups of order 3 are **isomorphic**.

(6) Observe that in the table we produced, $b = a \star a$. We will generally abbreviate $a \star a$ as a^2. So, another way to draw the table is as follows:

$\star$	e	a	a^2
e	e	a	a^2
a	a	a^2	e
a^2	a^2	e	a

This group is the **cyclic group of order 3**. We call it **cyclic** because the group consists of all powers of the single element a (the elements are a, a^2, and $a^3 = a^0 = e$). The **order** is the number of elements in the group.

6. Prove that in any monoid $(M, \star)$, the identity element is unique.

Proof: Let $(M, \star)$ be a monoid and suppose that e and f are both identity elements in M. Then, we have $f = e \star f = e$. Since we have shown f and e to be equal, there is only one identity element. $\square$

Notes: (1) The word "unique" means that there is only one. In mathematics, we often show that an object is unique by starting with two such objects and then arguing that they must actually be the same. Notice that in the proof above, when we said that e and f are both identity elements, we never insisted that they be *distinct* identity elements. And in fact, the end of the argument shows that they are not distinct.

(2) $e \star f = f$ because e is an identity element and $e \star f = e$ because f is an identity element.

7. Let $(F, +, \cdot)$ be a field. Prove each of the following:

   (i)   If $a, b \in F$ with $a + b = b$, then $a = 0$.

   (ii) If $a \in F$, $b \in F^*$, and $ab = b$, then $a = 1$.

   (iii) If $a \in F$, then $a \cdot 0 = 0$.

   (iv) If $a \in F^*$, $b \in F$, and $ab = 1$, then $b = \frac{1}{a}$.

   (v)   If $a, b \in F$ and $ab = 0$, then $a = 0$ or $b = 0$.

   (vi) If $a \in F$, then $-a = -1a$.

   (vii) $(-1)(-1) = 1$.

Proofs:

(i)   Let $a, b \in F$ with $a + b = b$. Then we have
$$a = a + 0 = a + \big(b + (-b)\big) = (a + b) + (-b) = b + (-b) = 0. \qquad \square$$

(ii) Let $a \in F$, $b \in F^*$, and $ab = b$. Then we have
$$a = a \cdot 1 = a(bb^{-1}) = (ab)b^{-1} = bb^{-1} = 1. \qquad \square$$

(iii) Let $a \in F$. Then $a \cdot 0 + a = a \cdot 0 + a \cdot 1 = a(0 + 1) = a \cdot 1 = a$. By (i), $a \cdot 0 = 0$. $\square$

(iv) Let $a \in F^*$, $b \in F$, and $ab = 1$. Then $b = 1b = (a^{-1}a)b = a^{-1}(ab) = a^{-1} \cdot 1 = a^{-1} = \frac{1}{a}$. $\square$

(v)   Let $a, b \in F$ and $ab = 0$. Assume that $a \neq 0$. Then $b = 1b = (a^{-1}a)b = a^{-1}(ab) = a^{-1} \cdot 0$. By (iii), $a^{-1} \cdot 0 = 0$. So, $b = 0$. $\square$

(vi) Let $a \in F$. Then $-1a + a = a(-1) + a \cdot 1 = a(-1 + 1) = a \cdot 0 = 0$ (by (iii)). So, $-1a$ is the additive inverse of a. Thus, $-1a = -a$. □

(vii) $(-1)(-1) + (-1) = (-1)(-1) + (-1) \cdot 1 = (-1)(-1 + 1) = (-1)(0) = 0$ (by (iii)). So, we see that $(-1)(-1)$ is the additive inverse of -1. Therefore, $(-1)(-1) = -(-1)$. □

8. Let $(F, +, \cdot)$ be a field with $\mathbb{N} \subseteq F$. Prove that $\mathbb{Q} \subseteq F$.

Proof: Let $n \in \mathbb{Z}$. If $n \in \mathbb{N}$, then $n \in F$ because $\mathbb{N} \subseteq F$. If $n \notin \mathbb{N}$, then $-n \in \mathbb{N}$. So, $-n \in F$. Since F is a field, we have $n = -(-n) \in F$. For each $n \in \mathbb{Z}^*$, $\frac{1}{n} = n^{-1} \in F$ because $n \in F$ and the multiplicative inverse property holds in F. Now, let $\frac{m}{n} \in \mathbb{Q}$. Then $m \in \mathbb{Z}$ and $n \in \mathbb{Z}^*$. Since $\mathbb{Z} \subseteq F$, $m \in F$. Since $n \in \mathbb{Z}^*$, we have $\frac{1}{n} \in F$. Therefore, $\frac{m}{n} = \frac{m \cdot 1}{1 \cdot n} = \frac{m}{1} \cdot \frac{1}{n} = m\left(\frac{1}{n}\right) \in F$ because F is closed under multiplication. Since $\frac{m}{n}$ was an arbitrary element of $\mathbb{Q}$, we see that $\mathbb{Q} \subseteq F$. □

LEVEL 3

9. Assume that a group $(G, \star)$ of order 4 exists with $G = \{e, a, b, c\}$, where e is the identity, $a^2 = b$ and $b^2 = e$. Construct the table for the operation of such a group.

Solution: Since $e \star x = x \star e = x$ for all x in the group, we can fill out the first row and the first column of the table as follows:

$\star$	e	a	b	c
e	e	a	b	c
a	a			
b	b			
c	c			

We now add in $a \star a = a^2 = b$ and $b \star b = b^2 = e$.

$\star$	e	a	b	c
e	e	a	b	c
a	a	b	$\boxdot$	
b	b		e	
c	c			

Now, the entry labeled with $\boxdot$ cannot be a or b because a and b appear in that row. It also cannot be e because e appears in that column. Therefore, the entry labeled with $\boxdot$ must be c. It follows that the entry to the right of $\boxdot$ must be e, and the entry at the bottom of the column must be a.

$\star$	e	a	b	c
e	e	a	b	c
a	a	b	c	e
b	b	$\odot$	e	
c	c		a	

Now, the entry labeled with $\odot$ cannot be b or e because b and e appear in that row. It also cannot be a because a appears in that column. Therefore, the entry labeled with $\odot$ must be c. The rest of the table is then determined.

$\star$	e	a	b	c
e	e	a	b	c
a	a	b	c	e
b	b	c	e	a
c	c	e	a	b

Note: Observe that in the table we produced, $b = a \star a = a^2$ and $c = b \star a = a^2 \star a = a^3$. So, another way to draw the table is as follows:

$\star$	e	a	a^2	a^3
e	e	a	a^2	a^3
a	a	a^2	a^3	e
a^2	a^2	a^3	e	a
a^3	a^3	e	a	a^2

This group is the **cyclic group of order 4**.

10. Prove that in any group $(G, \star)$, each element has a unique inverse.

Proof: Let $a \in G$ and suppose that $b, c \in G$ are both inverses of a. We will show that b and c must be the same. We have $c = c \star e = c \star (a \star b) = (c \star a) \star b = e \star b = b$. $\quad\square$

Notes: (1) $c = c \star e$ because e is an identity element.

(2) $e = a \star b$ because b is an inverse of a. So, $c \star e = c \star (a \star b)$.

(3) $c \star (a \star b) = (c \star a) \star b$ by associativity of $\star$.

(4) $c \star a = e$ because c is an inverse of a. So, $(c \star a) \star b = e \star b$.

(5) $e \star b = b$ because e is an identity element.

11. Let $(F, +, \cdot, \leq)$ be an ordered field. Prove each of the following:

    (i)   If $a, b \in F^+$ and $a > b$, then $\frac{1}{a} < \frac{1}{b}$.

    (ii)   If $a, b \in F$, then $a \geq b$ if and only if $-a \leq -b$.

    (iii)   If $a, b \in F^+$, then $a \leq b$ if and only if $a^2 \leq b^2$.

Proofs:

    (i)   Let $a, b \in F^+$ and $a > b$. Then $a - b = a + (-b) > b + (-b) = 0$ (we used Order Property (1) here). So, $\frac{1}{b} - \frac{1}{a} = \frac{1}{ab}(a - b)$. Since $a, b \in F^+$, $ab \in F^+$ by Order Property (2). So, $\frac{1}{ab} \in F^+$ by Theorem 6.8. Since $\frac{1}{ab} > 0$ and $a - b > 0$, we have $\frac{1}{b} - \frac{1}{a} = \frac{1}{ab}(a - b) > 0$. So, $\frac{1}{b} > \frac{1}{a}$, or equivalently, $\frac{1}{a} < \frac{1}{b}$. $\quad\square$

(ii) Let $a, b \in F$. Then $a \geq b$ if and only if $a - b \geq 0$ if and only if $-(a - b) \leq 0$ if and only if $-1\big(a + (-b)\big) \leq 0$ if and only if $-1a - 1(-b) \leq 0$ if and only if $-a - (-b) \leq 0$ if and only if $-a < -b$.
□

(iii) Let $a, b \in F^+$. By Order Property (1), $b + a \in F^+$. Therefore, $0 \leq b + a$. Also, we have $b^2 - a^2 = (b - a)(b + a)$.

If $a \leq b$, then $b - a \geq 0$. So, by Order Property (2), $0 \leq (b - a)(b + a) = b^2 - a^2$. Therefore, $a^2 \leq b^2$.

If $a \not\leq b$, then $b < a$, and so, $0 < a - b$. By Order Property (2), we have $0 < (a - b)(b + a) = a^2 - b^2$, and so, $b^2 < a^2$. Therefore, $a^2 \not\leq b^2$. By contrapositive, we have $b^2 < a^2$ implies $b < a$. So, $b^2 \leq a^2$ implies $b \leq a$.
□

12. Let $(F, +, \cdot)$ be a field. Show that $(F, \cdot)$ is a commutative monoid.

Proof: Let $(F, +, \cdot)$ be a field. Then $\cdot$ is a binary operation on F and $(F^*, \cdot)$ is a commutative group.

We first show that if $a \in F$, then $0a = 0$. To see this, observe that

$$0a + a = 0a + 1a = (0 + 1)a = 1a = a.$$

By Problem 7, part (i), $0a = 0$.

Let $x, y \in F$. If $x, y \in F^*$, then $xy = yx$. If $x = 0$, then $xy = 0y = 0$ by the previous result, and $yx = y \cdot 0 = 0$ by Problem 7, part (iii) above. If $y = 0$, then $xy = x \cdot 0 = 0$ by Problem 7, part (iii) above, and $yx = 0x = 0$ by the previous result. In all cases, we have $xy = yx$.

Next, let $x, y, z \in F$. If $x, y, z \in F^*$, then $(xy)z = x(yz)$. If $x = 0$, then by the previous result, we have $(xy)z = (0y)z = 0z = 0$ and $x(yz) = 0(yz) = 0$. If $y = 0$, by Problem 7, part (iii) and the previous result, we have $(xy)z = (x \cdot 0)z = 0z = 0$ and $x(yz) = x(0z) = x \cdot 0 = 0$. If $z = 0$, we have $(xy)z = (xy) \cdot 0 = 0$ and $x(yz) = x(y \cdot 0) = x \cdot 0 = 0$. In all cases, we have $(xy)z = x(yz)$.

Let $x \in F$. If $x \in F^*$, then $1x = x \cdot 1 = x$. If $x = 0$, then by Problem 7, part (iii), $1x = 1 \cdot 0 = 0$ and by the previous result, $x \cdot 1 = 0 \cdot 1 = 0$. In all cases, we have $1x = x \cdot 1 = x$.

Therefore, $(F, \cdot)$ is a commutative monoid.
□

LEVEL 4

13. Let $(G, \star)$ be a group with $a, b \in G$, and let a^{-1} and b^{-1} be the inverses of a and b, respectively. Prove

   (i) $(a \star b)^{-1} = b^{-1} \star a^{-1}$.

   (ii) the inverse of a^{-1} is a.

Proof of (i): Let $a, b \in G$. Then we have

$$(a \star b) \star (b^{-1} \star a^{-1}) = a \star \big(b \star (b^{-1} \star a^{-1})\big) = a \star \big((b \star b^{-1}) \star a^{-1}\big) = a \star (e \star a^{-1}) = a \star a^{-1} = e$$

and

$$(b^{-1} \star a^{-1}) \star (a \star b) = b^{-1} \star \big(a^{-1} \star (a \star b)\big) = b^{-1} \star \big((a^{-1} \star a) \star b\big) = b^{-1} \star (e \star b) = b^{-1} \star b = e.$$

So, $(a \star b)^{-1} = (b^{-1} \star a^{-1})$. □

Notes: (1) For the first and second equalities we used the associativity of $\star$ in G.

(2) For the third equality, we used the inverse property of $\star$ in G.

(3) For the fourth equality, we used the identity property of $\star$ in G.

(4) For the last equality, we again used the inverse property of $\star$ in G.

(5) Since multiplying $a \star b$ on either side by $b^{-1} \star a^{-1}$ results in the identity element e, it follows that $b^{-1} \star a^{-1}$ is the inverse of $a \star b$.

(6) In a group, to verify that an element h is the inverse of an element g, it suffices to show that $g \star h = e$ **or** $h \star g = e$. In other words, we can prove that $g \star h = e \to h \star g = e$ and we can prove that $h \star g = e \to g \star h = e$.

For a proof that $g \star h = e \to h \star g = e$, suppose that $g \star h = e$ and k is the inverse of g. Then $g \star k = k \star g = e$. Since $g \star h = e$ and $g \star k = e$, we have $g \star h = g \star k$. By multiplying by g^{-1} on each side of this equation, and using associativity, the inverse property, and the identity property, we get $h = k$. So, h is in fact the inverse of g.

Proving that $h \star g = e \to g \star h = e$ is similar. Thus, in the solution above, we need only show one of the sequences of equalities given. The second one follows for free.

Proof of (ii): Let $a \in G$. Since a^{-1} is the inverse of a, we have $a \star a^{-1} = a^{-1} \star a = e$. But this sequence of equations also says that a is the inverse of a^{-1}. □

14. Prove that there is no smallest positive real number.

Proof: Let $x \in \mathbb{R}^+$ and let $y = \frac{1}{2}x$. By Theorem 6.8, $\frac{1}{2} > 0$. So, by Order Property (2), $y > 0$.

Now, $x - y = x - \frac{1}{2}x = 1x - \frac{1}{2}x = \left(1 - \frac{1}{2}\right)x = \left(\frac{2}{2} - \frac{1}{2}\right)x = \frac{1}{2}x > 0$. So, $x > y$. It follows that y is a positive real number smaller than x. Since x was an arbitrary positive real number, there is no smallest positive real number. □

15. Let a be a nonnegative real number. Prove that $a = 0$ if and only if a is less than every positive real number. (Note: a nonnegative means $a \geq 0$.)

Proof: Let a be a nonnegative real number.

First suppose that $a = 0$. Let ϵ be a positive real number, so that $\epsilon > 0$. Then by direct substitution, $\epsilon > a$, or equivalently $a < \epsilon$. Since ϵ was an arbitrary positive real number, we have shown that a is less than every positive real number.

Now, suppose that a is less than every positive real number. Assume towards contradiction that $a \neq 0$. Then $a > 0$ (because a is nonnegative). Let $\epsilon = \frac{1}{2}a$. By the same reasoning used in Problem 14 above, we have that ϵ is a positive real number with $a > \epsilon$. This contradicts our assumption that a is less than every positive real number. $\qquad\square$

16. Prove that every rational number can be written in the form $\frac{m}{n}$, where $m \in \mathbb{Z}$, $n \in \mathbb{Z}^*$, and at least one of m or n is **not** even.

Proof: Let x be a rational number. Then there are $a \in \mathbb{Z}$ and $b \in \mathbb{Z}^*$ such that $x = \frac{a}{b}$. Let j be the largest integer such that 2^j divides a and let k be the largest integer such that 2^k divides b. Since, 2^j divides a, there is $c \in \mathbb{Z}$ such that $a = 2^j c$. Since, 2^k divides b, there is $d \in \mathbb{Z}$ such that $b = 2^k d$.

Observe that c is odd. Indeed, if c were even, then there would be an integer s such that $c = 2s$. But then $a = 2^j c = 2^j(2s) = (2^j \cdot 2)s = (2^j \cdot 2^1)s = 2^{j+1}s$. So, 2^{j+1} divides a, contradicting the maximality of j.

Similarly, d is odd.

So, we have $x = \frac{a}{b} = \frac{2^j c}{2^k d}$.

If $j \geq k$, then, $j - k \geq 0$ and $x = \frac{2^j c}{2^k d} = \frac{2^{j-k}c}{d}$. Let $m = 2^{j-k}c$ and $n = d$. Then $x = \frac{m}{n}$, $m \in \mathbb{Z}$ (because $\mathbb{Z}$ is closed under multiplication), $n \in \mathbb{Z}^*$ (if $n = 0$, then $b = 2^k d = 2^k n = 2^k \cdot 0 = 0$), and $n = d$ is odd.

If $j < k$, then $k - j > 0$ and $x = \frac{2^j c}{2^k d} = \frac{c}{2^{k-j}d}$. Let $m = c$ and $n = 2^{k-j}d$. Then $x = \frac{m}{n}$, $m = c \in \mathbb{Z}$, $n \in \mathbb{Z}^*$ (because $\mathbb{Z}$ is closed under multiplication, and if n were 0, then d would be 0, and then b would be 0), and $m = c$ is odd. $\qquad\square$

LEVEL 5

17. Prove that $(\mathbb{Q}, +, \cdot, \leq)$ and $(\mathbb{R}, +, \cdot, \leq)$ are ordered fields. Also prove that $(\mathbb{C}, +, \cdot)$ is field that cannot be ordered.

Proof: We first prove that $(\mathbb{Q}, +)$ is a commutative group.

(Closure) Let $x, y \in \mathbb{Q}$. Then there exist $a, c \in \mathbb{Z}$ and $b, d \in \mathbb{Z}^*$ such that $x = \frac{a}{b}$ and $y = \frac{c}{d}$. We have $x + y = \frac{a}{b} + \frac{c}{d} = \frac{ad+bc}{bd}$. Since $\mathbb{Z}$ is closed under multiplication, $ad \in \mathbb{Z}$ and $bc \in \mathbb{Z}$. Since $\mathbb{Z}$ is closed under addition, $ad + bc \in \mathbb{Z}$. Since $\mathbb{Z}^*$ is closed under multiplication, $bd \in \mathbb{Z}^*$. Therefore, $x + y \in \mathbb{Q}$.

(Associativity) Let $x, y, z \in \mathbb{Q}$. Then there exist $a, c, e \in \mathbb{Z}$ and $b, d, f \in \mathbb{Z}^*$ such that $x = \frac{a}{b}$, $y = \frac{c}{d}$, and $z = \frac{e}{f}$. Since multiplication and addition are associative in $\mathbb{Z}$, multiplication is (both left and right) distributive over addition in $\mathbb{Z}$, and multiplication is associative in $\mathbb{Z}^*$, we have

$$(x+y)+z = \left(\frac{a}{b}+\frac{c}{d}\right)+\frac{e}{f} = \frac{ad+bc}{bd}+\frac{e}{f} = \frac{(ad+bc)f+(bd)e}{(bd)f} = \frac{((ad)f+(bc)f)+(bd)e}{(bd)f}$$

$$= \frac{a(df)+(b(cf)+b(de))}{b(df)} = \frac{a(df)+b(cf+de)}{b(df)} = \frac{a}{b}+\frac{cf+de}{df} = \frac{a}{b}+\left(\frac{c}{d}+\frac{e}{f}\right) = x+(y+z).$$

(Identity) Let $\overline{0} = \frac{0}{1}$. We show that $\overline{0}$ is an identity for $(\mathbb{Q},+)$. Let $x \in \mathbb{Q}$. Then there exist $a \in \mathbb{Z}$ and $b \in \mathbb{Z}^*$ such that $x = \frac{a}{b}$. Since 0 is an identity for $\mathbb{Z}$, and $0 \cdot x = x \cdot 0 = 0$ for all $x \in \mathbb{Z}$, we have

$$x+\overline{0} = \frac{a}{b}+\frac{0}{1} = \frac{a \cdot 1 + b \cdot 0}{b \cdot 1} = \frac{a+0}{b} = \frac{a}{b} = x \quad \text{and} \quad \overline{0}+x = \frac{0}{1}+\frac{a}{b} = \frac{0b+1a}{1b} = \frac{0+a}{b} = \frac{a}{b} = x.$$

(Inverse) Let $x \in \mathbb{Q}$. Then there exist $a \in \mathbb{Z}$ and $b \in \mathbb{Z}^*$ such that $x = \frac{a}{b}$. Let $y = \frac{-1a}{b}$. Since $\mathbb{Z}$ is closed under multiplication, $-1a \in \mathbb{Z}$. So, $y \in \mathbb{Q}$. Since multiplication is associative and commutative in $\mathbb{Z}$ and $(-1)n = -n$ for all $n \in \mathbb{Z}$, we have

$$x+y = \frac{a}{b}+\frac{-1a}{b} = \frac{ab+b(-1a)}{b \cdot b} = \frac{ab+(-1a)b}{b^2} = \frac{ab+(-1)(ab)}{b^2} = \frac{ab-ab}{b^2} = \frac{0}{b^2} = \overline{0}$$

$$y+x = \frac{-1a}{b}+\frac{a}{b} = \frac{(-1a)b+ba}{b \cdot b} = \frac{-1(ab)+ab}{b^2} = \frac{-ab+ab}{b^2} = \frac{0}{b^2} = \overline{0}$$

So, y is the additive inverse of x.

(Commutativity) Let $x,y \in \mathbb{Q}$. Then there exist $a,c \in \mathbb{Z}$ and $b,d \in \mathbb{Z}^*$ such that $x = \frac{a}{b}$ and $y = \frac{c}{d}$. Since multiplication and addition are commutative in $\mathbb{Z}$, and multiplication is commutative in $\mathbb{Z}^*$, we have

$$x+y = \frac{a}{b}+\frac{c}{d} = \frac{ad+bc}{bd} = \frac{bc+ad}{db} = \frac{cb+da}{db} = \frac{c}{d}+\frac{a}{b} = y+x.$$

So, $(\mathbb{Q},+)$ is a commutative group.

We next prove that $(\mathbb{Q}^*, \cdot) = (\mathbb{Q} \setminus \{0\}, \cdot)$ is a commutative group.

(Closure) Let $x,y \in \mathbb{Q}^*$. Then there exist $a,b,c,d \in \mathbb{Z}^*$ such that $x = \frac{a}{b}$ and $y = \frac{c}{d}$. We have $xy = \frac{a}{b} \cdot \frac{c}{d} = \frac{ac}{bd}$. Since $\mathbb{Z}^*$ is closed under multiplication, $ac, bd \in \mathbb{Z}^*$. Therefore, $xy \in \mathbb{Q}^*$.

(Associativity) Let $x,y,z \in \mathbb{Q}^*$. Then there exist $a,b,c,d,e,f \in \mathbb{Z}^*$ such that $x = \frac{a}{b}$, $y = \frac{c}{d}$, and $z = \frac{e}{f}$. Since multiplication is associative in $\mathbb{Z}^*$, we have

$$(xy)z = \left(\frac{a}{b} \cdot \frac{c}{d}\right)\frac{e}{f} = \left(\frac{ac}{bd}\right)\frac{e}{f} = \frac{(ac)e}{(bd)f} = \frac{a(ce)}{b(df)} = \frac{a}{b}\left(\frac{ce}{df}\right) = \frac{a}{b}\left(\frac{c}{d} \cdot \frac{e}{f}\right) = x(yz).$$

(Identity) Let $\overline{1} = \frac{1}{1}$. We show that $\overline{1}$ is an identity for $(\mathbb{Q}^*, \cdot)$. Let $x \in \mathbb{Q}^*$. Then there exist $a,b \in \mathbb{Z}^*$ such that $x = \frac{a}{b}$. Since 1 is an identity for $\mathbb{Z}^*$, we have

$$x \cdot \overline{1} = \frac{a}{b} \cdot \frac{1}{1} = \frac{a \cdot 1}{b \cdot 1} = \frac{a}{b} = x \text{ and } \overline{1}x = \frac{1}{1} \cdot \frac{a}{b} = \frac{1a}{1b} = \frac{a}{b} = x.$$

(Inverse) Let $x \in \mathbb{Q}^*$. Then there exist $a, b \in \mathbb{Z}^*$ such that $x = \frac{a}{b}$. Let $y = \frac{b}{a}$. Then $y \in \mathbb{Q}^*$ (note that $a \neq 0$). Since multiplication is commutative in $\mathbb{Z}^*$, we have

$$xy = \frac{a}{b} \cdot \frac{b}{a} = \frac{ab}{ba} = \frac{ab}{ab} = \frac{1}{1} = \overline{1}.$$

So, y is the multiplicative inverse of x.

(Commutativity) Let $x, y \in \mathbb{Q}^*$. Then there exist $a, b, c, d \in \mathbb{Z}^*$ such that $x = \frac{a}{b}$ and $y = \frac{c}{d}$. Since multiplication is commutative in $\mathbb{Z}^*$, we have

$$xy = \frac{a}{b} \cdot \frac{c}{d} = \frac{ac}{bd} = \frac{ca}{db} = \frac{c}{d} \cdot \frac{a}{b} = yx.$$

So, $(\mathbb{Q}^*, \cdot)$ is a commutative group.

Now we prove that multiplication is distributive over addition in $\mathbb{Q}$.

(Distributivity) Let $x, y, z \in \mathbb{Q}$. Then there exist $a, c, e \in \mathbb{Z}$ and $b, d, f \in \mathbb{Z}^*$ such that $x = \frac{a}{b}, y = \frac{c}{d}$, and $z = \frac{e}{f}$. Let's start with left distributivity.

$$x(y + z) = \frac{a}{b}\left(\frac{c}{d} + \frac{e}{f}\right) = \frac{a}{b}\left(\frac{cf + de}{df}\right) = \frac{a(cf + de)}{b(df)}$$

$$xy + xz = \frac{a}{b} \cdot \frac{c}{d} + \frac{a}{b} \cdot \frac{e}{f} = \frac{ac}{bd} + \frac{ae}{bf} = \frac{(ac)(bf) + (bd)(ae)}{(bd)(bf)}$$

We need to verify that $\frac{(ac)(bf)+(bd)(ae)}{(bd)(bf)} = \frac{a(cf+de)}{b(df)}$.

Since $\mathbb{Z}$ is a ring, $(ac)(bf) + (bd)(ae) = bacf + bade = ba(cf + de)$ (see Note 1 below).

Since multiplication is associative and commutative in $\mathbb{Z}^*$, we have

$$(bd)(bf) = b(d(bf)) = b((db)f) = b((bd)f) = b(b(df)).$$

So, $\frac{(ac)(bf)+(bd)(ae)}{(bd)(bf)} = \frac{ba(cf+de)}{b(b(df))} = \frac{a(cf+de)}{b(df)}$.

For right distributivity, we can use left distributivity together with the commutativity of multiplication in $\mathbb{Q}$.

$$(y + z)x = x(y + z) = xy + xz = yx + zx \qquad \square$$

Notes: (1) We skipped many steps when verifying $(ac)(bf) + (bd)(ae) = ba(cf + de)$. The dedicated reader may want to verify this equality carefully, making sure to use only the fact that $\mathbb{Z}$ is a ring, and making a note of which ring property is being used at each step.

(2) In the very last step of the proof, we cancelled one b in the numerator of the fraction with b in the denominator of the fraction. In general, if $j \in \mathbb{Z}$ and $m, k \in \mathbb{Z}^*$, then $\frac{mj}{mk} = \frac{j}{k}$. To verify that this is true, simply observe that since $\mathbb{Z}$ is a ring, we have $(mj)k = m(jk) = m(kj) = (mk)j$.

We next prove the order properties. Let $\frac{a}{b}, \frac{c}{d}, \frac{e}{f} \in \mathbb{Q}$ and assume that $\frac{a}{b} \leq \frac{c}{d}$. Then $ad \leq bc$. Since $\mathbb{Z}$ is an ordered ring, we get

$$(af + be)(df) = afdf + bedf = adff + bedf \leq bcff + bedf = bf(cf + de).$$

Therefore, $\frac{a}{b} + \frac{e}{f} = \frac{af+be}{bf} \leq \frac{cf+de}{df} = \frac{c}{d} + \frac{e}{f}$.

Next, let $\frac{a}{b}, \frac{c}{d} \in \mathbb{Q}$ with $0 \leq \frac{a}{b}$ and $0 \leq \frac{c}{d}$. We may assume that $a, b, c, d \geq 0$. Then $\frac{a}{b} \cdot \frac{c}{d} = \frac{ac}{bd} \geq 0$.

It follows that $(\mathbb{Q}, +, \cdot, \leq)$ is an ordered field.

We proved that $(\mathbb{R}, +)$ is a commutative group in part 6 of Example 6.3. We next prove that $(\mathbb{R}^*, \cdot) = (\mathbb{R} \setminus \{0\}, \cdot)$ is a commutative group.

(Closure) This is Problem 19 from Problem Set 4.

(Associativity) To see that $\cdot$ is associative in $\mathbb{R}^*$, we use the associativity of $\cdot$ in $\mathbb{Q}$. If $[(x_n)], [(y_n)], [(z_n)] \in \mathbb{R}^*$, then

$$([(x_n)] \cdot [(y_n)]) \cdot [(z_n)] = [(x_n \cdot y_n)] \cdot [(z_n)] = [((x_n \cdot y_n) \cdot z_n)]$$
$$= [(x_n \cdot (y_n \cdot z_n))] = [x_n] \cdot [(y_n \cdot z_n)] = [(x_n)] \cdot ([(y_n)] \cdot [(z_n)]).$$

(Identity) To see that $[(1)]$ is the multiplicative identity, using the fact that 1 is the additive identity in $\mathbb{Q}$, we have for $[(x_n)] \in \mathbb{R}$,

$$[(1)] \cdot [(x_n)] = [(1 \cdot x_n)] = [(x_n)] \text{ and } [(x_n)] \cdot [(1)] = [(x_n \cdot 1)] = [(x_n)].$$

(Inverse) The inverse of the real number $[(x_n)]$ is $[(y_n)]$, where for each $n \in \mathbb{N}$, $y_n = \frac{1}{x_n}$ if $x_n \neq 0$ and $y_n = 0$ if $x_n = 0$. We have that $[(x_n)] \cdot [(y_n)] = [(z_n)]$ and $[(y_n)] \cdot [(x_n)] = [(z_n)]$, where z_n is 0 or 1 for all $n \in \mathbb{N}$. We claim that $[(z_n)] = [(1)]$. To see this, note that since $[(x_n)] \neq [(0)]$, there is a $K > 0$ such that for $n > N$, $x_n \neq 0$.

(Commutativity) To see that $\cdot$ is commutative in $\mathbb{R}^*$, we use the commutativity of $\cdot$ in $\mathbb{Q}$. If $[(x_n)], [(y_n)] \in \mathbb{R}^*$, then $[(x_n)] \cdot [(y_n)] = [(x_n y_n)] = [(y_n x_n)] = [(y_n)] \cdot [(x_n)]$.

(Distributivity) Distributivity is similar to commutativity and associativity.

We next prove the order properties. Let $[(x_n)], [(y_n)], [(z_n)] \in \mathbb{R}$ and assume that $[(x_n)] \leq [(y_n)]$. Then there is $K \in \mathbb{N}$ such that $n > K$ implies $x_n \leq y_n$. Since $\mathbb{Q}$ is an ordered ring, it follows that $n > K$ implies $x_n + z_n \leq y_n + z_n$. So, $[(x_n)] + [(z_n)] = [(x_n + z_n)] \leq [(y_n + z_n)] = [(y_n)] + [(z_n)]$.

Next, let $[(x_n)], [(y_n)] \in \mathbb{R}$ with $[(0)] \leq [(x_n)]$ and $[(0)] \leq [(y_n)]$. Then there is $K_1 \in \mathbb{N}$ such that $n > K_1$ implies $x_n \geq 0$ and there is $K_2 \in \mathbb{N}$ such that $n > K_2$ implies $y_n \geq 0$. Let $K = \max\{K_1, K_2\}$ and let $n > K$. Since $K \geq K_1$, $x_n \geq 0$. Since $K \geq K_2$, $y_n \geq 0$. Since $\mathbb{Q}$ is an ordered ring, $x_n \cdot y_n \geq 0$. Therefore, we have $[(x_n)] \cdot [(y_n)] \geq [(0)]$.

It follows that $(\mathbb{R}, +, \cdot, \leq)$ is an ordered field.

We now prove that $(\mathbb{C}, +)$ is a commutative group.

(Closure) Let $z, w \in \mathbb{C}$. Then there are $a, b, c, d \in \mathbb{R}$ such that $z = a + bi$ and $w = c + di$. By definition, $z + w = (a + bi) + (c + di) = (a + c) + (b + d)i$. Since $\mathbb{R}$ is closed under addition, $a + b \in \mathbb{R}$ and $c + d \in \mathbb{R}$. Therefore, $z + w \in \mathbb{C}$.

(Associativity) Let $z, w, v \in \mathbb{C}$. Then there are $a, b, c, d, e, f \in \mathbb{R}$ such that $z = a + bi$, $w = c + di$, and $v = e + fi$. Since addition is associative in $\mathbb{R}$, we have

$$(z + w) + v = \big((a + bi) + (c + di)\big) + (e + fi) = \big((a + c) + (b + d)i\big) + (e + fi)$$
$$= \big((a + c) + e\big) + \big((b + d) + f\big)i = \big(a + (c + e)\big) + \big(b + (d + f)\big)i$$
$$= (a + bi) + \big((c + e) + (d + f)i\big) = (a + bi) + \big((c + di) + (e + fi)\big) = z + (w + v).$$

(Commutativity) Let $z, w \in \mathbb{C}$. Then there are $a, b, c, d \in \mathbb{R}$ such that $z = a + bi$ and $w = c + di$. Since addition is commutative in $\mathbb{R}$, we have

$$z + w = (a + bi) + (c + di) = (a + c) + (b + d)i = (c + a) + (d + b)i$$
$$= (c + di) + (a + bi) = w + z.$$

(Identity) Let $\overline{0} = 0 + 0i$. We show that $\overline{0}$ is an additive identity for $\mathbb{C}$. Since $0 \in \mathbb{R}$, $\overline{0} \in \mathbb{C}$. Let $z \in \mathbb{C}$. Then there are $a, b \in \mathbb{R}$ such that $z = a + bi$. Since 0 is an additive identity in $\mathbb{R}$, we have

$$\overline{0} + z = (0 + 0i) + (a + bi) = (0 + a) + (0 + b)i = a + bi.$$
$$z + \overline{0} = (a + bi) + (0 + 0i) = (a + 0) + (b + 0)i = a + bi.$$

(Inverse) Let $z \in \mathbb{C}$. Then there are $a, b \in \mathbb{R}$ such that $z = a + bi$. Let $w = -a + (-b)i$. Then

$$z + w = (a + bi) + (-a + (-b)i) = \big(a + (-a)\big) + \big(b + (-b)\big)i = 0 + 0i = \overline{0}.$$
$$w + z = (-a + (-b)i) + (a + bi) = (-a + a) + (-b + b)i = 0 + 0i = \overline{0}.$$

We next prove that $(\mathbb{C}^*, \cdot)$ is a commutative group.

(Closure) Let $z, w \in \mathbb{C}^*$. Then there are $a, b, c, d \in \mathbb{R}$ such that $z = a + bi$ and $w = c + di$. By definition, $zw = (a + bi)(c + di) = (ac - bd) + (ad + bc)i$. Since $\mathbb{R}$ is closed under multiplication, we have $ac, bd, ad, bc \in \mathbb{R}$. Also, $-bd$ is the additive inverse of bd in $\mathbb{R}$. Since $\mathbb{R}$ is closed under addition, we have $ac - bd = ac + (-bd) \in \mathbb{R}$ and $ad + bc \in \mathbb{R}$. Therefore, $zw \in \mathbb{C}$.

We still need to show that $zw \neq 0$. If $zw = 0$, then $ac - bd = 0$ and $ad + bc = 0$. So, $ac = bd$ and $ad = -bc$. Multiplying each side of the last equation by c gives us $acd = -bc^2$. Replacing ac with bd on the left gives $bd^2 = -bc^2$, or equivalently, $bd^2 + bc^2 = 0$. So, $b(d^2 + c^2) = 0$. If $d^2 + c^2 = 0$, then $c = 0$ and $d = 0$, and so, $w = 0$. If $b = 0$, then $ac = 0$, and so, $a = 0$ or $c = 0$. If $a = 0$, then $z = 0$. If $c = 0$ and $a \neq 0$, then since $ad = -bc = 0$, we have $d = 0$. So, $w = 0$. So, we see that $zw = 0$ implies $z = 0$ or $w = 0$. By contrapositive, since $z, w \in \mathbb{C}^*$, we must have $zw \neq 0$, and so, $zw \in \mathbb{C}^*$.

(Associativity) Let $z, w, v \in \mathbb{C}^*$. Then there are $a, b, c, d, e, f \in \mathbb{R}$ such that $z = a + bi$, $w = c + di$, and $v = e + fi$. Since addition and multiplication are associative in $\mathbb{R}$, addition is commutative in $\mathbb{R}$, and multiplication is distributive over addition in $\mathbb{R}$, we have

$$(zw)v = \big((a+bi)(c+di)\big)(e+fi) = \big((ac-bd)+(ad+bc)i\big)(e+fi)$$
$$= [(ac-bd)e - (ad+bc)f] + [(ac-bd)f + (ad+bc)e]i$$
$$= (ace - bde - adf - bcf) + (acf - bdf + ade + bce)i$$
$$= (ace - adf - bcf - bde) + (acf + ade + bce - bdf)i$$
$$= [a(ce-df) - b(cf+de)] + [a(cf+de) + b(ce-df)]i$$
$$= (a+bi)\big((ce-df) + (cf+de)i\big) = (a+bi)\big((c+di)(e+fi)\big) = z(wv).$$

(Commutativity) Let $z, w \in \mathbb{C}^*$. Then there are $a, b, c, d \in \mathbb{R}$ such that $z = a + bi$ and $w = c + di$. Since addition and multiplication are commutative in $\mathbb{R}$, we have

$$zw = (a+bi)(c+di) = (ac-bd) + (ad+bc)i$$
$$= (ca-db) + (cb+da)i = (c+di)(a+bi) = wz$$

(Identity) Let $\overline{1} = 1 + 0i$. We show that $\overline{1}$ is a multiplicative identity for $\mathbb{C}^*$. Since $0, 1 \in \mathbb{R}$, $\overline{1} \in \mathbb{C}^*$. Let $z \in \mathbb{C}^*$. Then there are $a, b \in \mathbb{R}$ such that $z = a + bi$. Since 0 is an additive identity in $\mathbb{R}$, 1 is a multiplicative identity in $\mathbb{R}$, and $0 \cdot x = x \cdot 0 = 0$ for all $x \in \mathbb{R}$, we have

$$\overline{1}z = (1+0i)(a+bi) = (1a - 0b) + (1b + 0a)i = 1a + 1bi = a + bi.$$

$$z \cdot \overline{1} = (a+bi)(1+0i) = (a \cdot 1 - b \cdot 0) + (a \cdot 0 + b \cdot 1)i = a \cdot 1 + b \cdot 1i = a + bi.$$

(Inverse) Let $z \in \mathbb{C}^*$. Then there are $a, b \in \mathbb{R}$ such that $z = a + bi$. Let $w = \frac{a}{a^2+b^2} + \frac{-b}{a^2+b^2}i$. Then we have

$$zw = (a+bi)\left(\frac{a}{a^2+b^2} + \frac{-b}{a^2+b^2}i\right)$$

$$= \left(a \cdot \frac{a}{a^2+b^2} - b \cdot \frac{-b}{a^2+b^2}\right) + \left(a \cdot \frac{-b}{a^2+b^2} + b \cdot \frac{a}{a^2+b^2}\right)i$$

$$= \frac{a^2+b^2}{a^2+b^2} + \frac{-ab+ba}{a^2+b^2}i = 1 + 0i = \overline{1}.$$

$$wz = \left(\frac{a}{a^2+b^2} + \frac{-b}{a^2+b^2}i\right)(a+bi)$$

$$= \left(\frac{a}{a^2+b^2} \cdot a - \frac{-b}{a^2+b^2} \cdot b\right) + \left(\frac{a}{a^2+b^2} \cdot b + \frac{-b}{a^2+b^2} \cdot a\right)i$$

$$= \frac{a^2+b^2}{a^2+b^2} + \frac{ab-ba}{a^2+b^2}i = 1 + 0i = \overline{1}.$$

(Left Distributivity) Let $z, w, v \in \mathbb{C}$. Then there are $a, b, c, d, e, f \in \mathbb{R}$ such that $z = a + bi$, $w = c + di$, and $v = e + fi$. Since multiplication is left distributive over addition in $\mathbb{R}$, and addition is associative and commutative in $\mathbb{R}$, we have

$$z(w + v) = (a + bi)[(c + di) + (e + fi)] = (a + bi)[(c + e) + (d + f)i]$$
$$= [a(c + e) - b(d + f)] + [a(d + f) + b(c + e)]i$$
$$= (ac + ae - bd - bf) + (ad + af + bc + be)i$$
$$= [(ac - bd) + (ad + bc)i\,] + [(ae - bf) + (af + be)i]$$
$$(a + bi)(c + di) + (a + bi)(e + fi) = zw + zv.$$

(Right Distributivity) Let $z, w, v \in \mathbb{C}$. There are $a, b, c, d, e, f \in \mathbb{R}$ such that $z = a + bi$, $w = c + di$, and $v = e + fi$. Since multiplication is right distributive over addition in $\mathbb{R}$, and addition is associative and commutative in $\mathbb{R}$, we have

$$(w + v)z = [(c + di) + (e + fi)](a + bi) = [(c + e) + (d + f)i](a + bi)$$
$$= [(c + e)a - (d + f)b] + [(c + e)b + (d + f)a]i$$
$$= (ca + ea - db - fb) + (cb + eb + da + fa)i$$
$$= [(ca - db) + (cb + da)i\,] + [(ea - fb) + (eb + fa)i]$$
$$(c + di)(a + bi) + (e + fi)(a + bi) = wz + vz.$$

Therefore, $(\mathbb{C}, +, \cdot)$ is field.

By part 3 of Example 6.5, $(\mathbb{C}, +, \cdot)$ cannot be ordered. $\qquad\qquad$ □

Note: When verifying the inverse property for multiplication, we didn't mention the field properties that were used and we skipped some steps. The dedicated reader may want to fill in these details.

> 18. Prove that every nonempty set of real numbers that is bounded below has a greatest lower bound in $\mathbb{R}$.

Proof: Let S be a nonempty set of real numbers that is bounded below. Let K be a lower bound of S, so that for all $x \in S$, $x \geq K$. Define the set T by $T = \{-x \mid x \in S\}$.

Let $y \in T$. Then there is $x \in S$ with $y = -x$. Since $x \in S$, $x \geq K$. It follows from Problem 11, part (ii) that $y = -x \leq -K$. Since $y \in T$ was arbitrary, we have shown that for all $y \in T$, $y \leq -K$. It follows that $-K$ is an upper bound of the set T.

By the Completeness Property of $\mathbb{R}$, T has a least upper bound M. We will show that $-M$ is a greatest lower bound of S.

Let $x \in S$. Then $-x \in T$. Since M is an upper bound of T, $-x \leq M$. So, by Problem 11, part (ii), $x \geq -M$. Since $x \in S$ was arbitrary, we have shown that for all $x \in S$, $x \geq -M$. Therefore, $-M$ is a lower bound of S.

Let $B > -M$. Then $-B < M$. Since M is the least upper bound of T, there is $y \in T$ with $y > -B$. By Problem 11, part (ii), we have $-y < B$. Since $y \in T$, $-y \in S$. Thus, B is not a lower bound of S.

Therefore, $-M$ is a greatest lower bound of S.

Since S was arbitrary, we have shown that every nonempty set of real numbers that is bounded below has a greatest lower bound in $\mathbb{R}$. □

19. Show that between any two real numbers there is a real number that is **not** rational.

Proof: Let $x, y \in \mathbb{R}$ with $x < y$. Let c be a positive number that is not rational. Then $\frac{x}{c} < \frac{y}{c}$. By the Density Theorem, there is a $q \in \mathbb{Q}$ such that $\frac{x}{c} < q < \frac{y}{c}$. We can assume that $q \neq 0$ (if it were, we could simply apply the Density Theorem again to get $p \in \mathbb{Q}$ with $\frac{x}{c} < p < q$, and p would not be 0). It follows that $x < cq < y$. Since $c = (cq)q^{-1}$, it follows that $cq \notin \mathbb{Q}$ (if $cq \in \mathbb{Q}$, then $c \in \mathbb{Q}$ because $\mathbb{Q}$ is closed under multiplication). So, cq is a real number between x and y that is **not** rational. □

20. Let $T = \{x \in F \mid -2 < x \leq 2\}$. Prove $\sup T = 2$ and $\inf T = -2$.

Proof: If $x \in T$, then by the definition of T, $x \leq 2$. So, 2 is an upper bound of T.

Now, let $B < 2$, and let $z = \max\left\{0, \frac{1}{2}(B + 2)\right\}$. Since $B < 2$, we have

$$\frac{1}{2}(B + 2) < \frac{1}{2}(2 + 2) = \frac{1}{2} \cdot 4 = 2.$$

So, if we have $\frac{1}{2}(B + 2) > 0$, then $\frac{1}{2}(B + 2) \in T$. Since $0 \in T$, we see that $z \in T$. Also,

$$z \geq \frac{1}{2}(B + 2) > \frac{1}{2}(B + B) = \frac{1}{2}(2B) = \left(\frac{1}{2} \cdot 2\right)B = 1B = B.$$

So, we see that $z \in T$ and $z > B$. Therefore, B is not an upper bound of T. So, $2 = \sup T$.

If $x \in T$, then by the definition of T, $x > -2$. So, -2 is a lower bound of T.

Now, let $C > -2$, and let $w = \min\left\{0, \frac{1}{2}(-2 + C)\right\}$. Since $C > -2$, we have

$$\frac{1}{2}(-2 + C) > \frac{1}{2}(-2 - 2) = \frac{1}{2}(-4) = -2.$$

So, if we have $\frac{1}{2}(-2 + C) < 0$, then $\frac{1}{2}(-2 + C) \in T$. Since $0 \in T$, we see that $w \in T$. Also,

$$w \leq \frac{1}{2}(-2 + C) < \frac{1}{2}(C + C) = \frac{1}{2}(2C) = \left(\frac{1}{2} \cdot 2\right)C = 1C = C.$$

So, we see that $w \in T$ and $w < C$. Therefore, C is not a lower bound of T. So, $-2 = \inf T$. □

Problem Set 7

LEVEL 1

1. Define a set of real numbers with exactly two accumulation points.

Solution: Let $S = \left\{ (-1)^n \left(1 + \frac{1}{n} \right) \mid n \in \mathbb{N} \right\}$. Then S has exactly two accumulation points: 1 and -1.

2. Let I be an open or half-open interval of real numbers. Prove that I is **not** compact.

Proof: Let $I = (a, b)$. The collection of open sets $\mathcal{C} = \left\{ \left(a + \frac{1}{n}, b \right) \mid n \in \mathbb{Z}^+ \right\}$ is an open covering of I with no finite subcover.

Let $J = (a, b]$. The collection $\mathcal{C} = \left\{ \left(a + \frac{1}{n}, b + 1 \right) \mid n \in \mathbb{Z}^+ \right\}$ is an open covering of J with no finite subcover.

Let $K = [a, b)$. The collection $\mathcal{C} = \left\{ \left(a - 1, b - \frac{1}{n} \right) \mid n \in \mathbb{Z}^+ \right\}$ is an open covering of K with no finite subcover. $\square$

3. Prove that a finite union of compact subsets of $\mathbb{R}$ is compact.

Proof: For each $k = 0, 1, \ldots, n$, let A_k be a compact subset of $\mathbb{R}$. Let $A = \bigcup \{ A_k \mid k = 0, 1, \ldots, n \}$. We will show that A is compact. To this end, let $\mathcal{C}$ be an open covering of A. Then for each $k = 0, 1, \ldots, n$, $\mathcal{C}$ is an open covering of A_k. Since A_k is compact, there is a finite subcollection $\mathcal{C}_k$ such that $\mathcal{C}_k$ still covers A_k. Then $\bigcup \{ \mathcal{C}_k \mid k = 0, 1, \ldots, n \}$ is a finite subcollection of $\mathcal{C}$ that covers A. $\square$

4. Prove that the Cantor set is a compact subset of $\mathbb{R}$.

Proof: Let C be the Cantor set. We saw in part 2 of Example 7.22 that C is closed. Since $C \subseteq [0, 1]$, C is bounded. So, by the Heine-Borel Theorem (Theorem 7.28), C is compact. $\square$

LEVEL 2

5. Let S be a set of real numbers and let S' be the set of accumulation points of S. Prove that S' is closed in $\mathbb{R}$.

Proof: Let S be a set of real numbers, let S' be the set of accumulation points of S, and let x be an accumulation point of S'. We will show that x is an accumulation point of S. To see this, let (a, b) be an open interval containing x. Since x is an accumulation point of S', there is $y \in S'$ with $y \in (a, b)$ and $y \neq x$. Without loss of generality, assume that $x < y$. Since $y \in S'$, y is an accumulation point of S. So, there is $z \in S$ with $z \in (x, b)$. Thus, we have found an element of S in (a, b) different from x. So, x is an accumulation point of S, and therefore, $x \in S'$. Since x was an arbitrary accumulation point of S', it follows that S' contains each of its accumulation points. By Theorem 7.24, S' is closed in $\mathbb{R}$. $\square$

6. Determine the accumulation points of each of the following subsets of $\mathbb{R}$:

   (i) $\{(-1)^n \mid n \in \mathbb{Z}^+\}$

   (ii) $\{x \mid |x| < 1\}$

   (iii) $\{x \mid 0 < |x - 2| \leq 3\}$

Solutions:

(i) This set is equal to $\{1, -1\}$. It has **no accumulation points**.

(ii) The set of accumulation points of the set $\{x \mid |x| < 1\}$ is the set $\{x \mid |x| \leq 1\}$.

(iii) The set of accumulation points of the set $\{z \mid 0 < |x - 2| \leq 3\}$ is the set $\{x \mid |x - 2| \leq 3\}$.

7. Must the union of countably many compact subsets of $\mathbb{R}$ be compact? If so, provide a proof. If not, provide a counterexample.

Solution: No. Let $X = \{[n, n + 1] \mid n \in \mathbb{Z}\}$. Each set in X is compact, but $\bigcup X = \mathbb{R}$, which is not compact.

LEVEL 3

8. Prove the following:

   (i) For all $b \in \mathbb{R}$, the infinite interval $(-\infty, b)$ is open in $\mathbb{R}$.

   (ii) The intersection of two open intervals in $\mathbb{R}$ is either empty or an open interval in $\mathbb{R}$.

   (iii) The intersection of finitely many open sets in $\mathbb{R}$ is open in $\mathbb{R}$.

Proofs:

(i) Let $x \in (-\infty, b)$ and let $a = x - 1$. Since $x \in (-\infty, b)$, $x < b$. Since $x - (x - 1) = 1 > 0$, we have $x > x - 1 = a$. So, we have $a < x < b$. That is, $x \in (a, b)$. Also, $(a, b) \subseteq (-\infty, b)$. Since $x \in (-\infty, b)$ was arbitrary, $(-\infty, b)$ is an open set. □

(ii) Let (a, b) and (c, d) be open intervals in $\mathbb{R}$ (a and c can be $-\infty$, and b and d can be ∞, where $-\infty$ is less than any real number and ∞, and ∞ is greater than any real number and $-\infty$). Without loss of generality, we may assume that $a \leq c$. If $b \leq c$, then we have $(a, b) \cap (c, d) = \emptyset$ because if $a < x < b$ and $c < x < d$, then $x < b \leq c < x$, and so, $x < x$, which is impossible.

So, we may assume that $c < b$. Let $e = \min\{b, d\}$. We claim that $(a, b) \cap (c, d) = (c, e)$.

Let $x \in (a, b) \cap (c, d)$. Then $x \in (a, b)$ and $x \in (c, d)$. So, $a < x < b$ and $c < x < d$. In particular, $x > c$, $x < b$, and $x < d$. Since $x < b$ and $x < d$, $x < e$. So, $x \in (c, e)$. Since $x \in (a, b) \cap (c, d)$ was arbitrary, we have shown that $(a, b) \cap (c, d) \subseteq (c, e)$.

Now, let $x \in (c, e)$. Then $c < x < e$. We are assuming that $a \leq c$. We also have $e \leq b$. So, $a \leq c < x < e \leq b$. Therefore, $x \in (a, b)$. We also have $e \leq d$. So, $c < x < e \leq d$, and therefore, $x \in (c, d)$. Since $x \in (a, b)$ and $x \in (c, d)$, we have $x \in (a, b) \cap (c, d)$. Since $x \in (c, e)$ was arbitrary, we have shown that $(c, e) \subseteq (a, b) \cap (c, d)$.

Finally, since we have shown $(a, b) \cap (c, d) \subseteq (c, e)$ and $(c, e) \subseteq (a, b) \cap (c, d)$, we have $(a, b) \cap (c, d) = (c, e)$.

Therefore, the intersection of two open intervals in $\mathbb{R}$ is either empty or an open interval in $\mathbb{R}$. $\qquad \square$

(iii)   The intersection of a single set with itself is just that set itself, and so, the result holds trivially for one open set.

So, we will prove the following statement: "The intersection of a set of finitely many open sets in $\mathbb{R}$ consisting of at least 2 sets is an open set in $\mathbb{R}$." We will prove this by induction on the number of open sets we are taking the intersection of. Theorem 7.19 is the base case $n = 2$.

For the inductive step, assume that the intersection of k nonempty open sets in $\mathbb{R}$ is open, and let X be a set of $k + 1$ open sets. Let $A \in X$ and let B be the intersection of all the sets in X except A. By the induction hypotheses, B is open. Therefore, $\cap X = A \cap B$ is open by Theorem 7.19.

By the Principle of Mathematical Induction, we have shown that the intersection of a set of finitely many open sets in $\mathbb{R}$ consisting of at least 2 sets is an open set in $\mathbb{R}$. $\qquad \square$

9. Prove the Triangle Inequality (Theorem 7.4).

Proof: $|x + y|^2 = (x + y)(x + y) = x^2 + 2xy + y^2 \leq |x|^2 + 2|x||y| + |y|^2 = (|x| + |y|)^2$. Since $|x + y|$ and $|x| + |y|$ are nonnegative, by part (iii) of Problem 11 from Problem Set 6, we have $|x + y| \leq |x| + |y|$. $\qquad \square$

10. Let x and y be real numbers. Prove $\bigl||x| - |y|\bigr| \leq |x \pm y| \leq |x| + |y|$.

Proof: $|x| = |(x + y) + (-y)| \leq |x + y| + |-y| = |x + y| + |y|$. So, $|x + y| \geq |x| - |y|$.

$|y| = |(x + y) + (-x)| \leq |x + y| + |-x| = |x + y| + |x|$. So, $|x + y| \geq |y| - |x| = -(|x| - |y|)$.

Since for all $x, y \in \mathbb{R}$, we have $\bigl||x| - |y|\bigr| = |x| - |y|$ or $\bigl||x| - |y|\bigr| = -(|x| - |y|)$, it follows that $\bigl||x| - |y|\bigr| \leq |x + y|$.

Combining this result with the Triangle Inequality, gives us $\bigl||x| - |y|\bigr| \leq |x + y| \leq |x| + |y|$.

Now, by the Triangle Inequality we have $|x - y| = |x + (-y)| \leq |x| + |-y| = |x| + |y|$.

Finally, by the third paragraph, we have $|x - y| = |x + (-y)| \geq \bigl||x| - |-y|\bigr| = \bigl||x| - |y|\bigr|$. $\qquad \square$

11. Let $x \in \mathbb{R}$ be an accumulation point of a set S. Prove that every open interval containing x contains infinitely many points of S.

Proof: Let $x \in \mathbb{R}$ be an accumulation point of a set S, let (a, b) be an open interval containing x, and let $X = \{x_1, x_2, \ldots, x_n\}$ be an arbitrary subset of $(a, b) \cap S \cap (\mathbb{R} \setminus \{x\})$. Let $A = \{x_i \mid x_i < x\} \cup \{a\}$ and let $B = \{x_i \mid x_i > x\} \cup \{b\}$. Now, let c be the largest element of A and let d be the smallest element of B. Then $c < x < d$, and so, (c, d) is an open interval containing x. Since x is an accumulation point of S, there is $y \in (c, d)$ with $y \in S$ and $y \neq x$. Since $a \leq c$ and $d \leq b$, $(c, d) \subseteq (a, b)$. Therefore, $y \in (a, b)$. Also, $y \notin X$. So, given any finite collection of elements of $Y = (a, b) \cap S \cap (\mathbb{R} \setminus \{x\})$, we can always find another element of Y that is not in that finite collection. $\square$

12. Let X, Y, and Z be sets of real numbers with $Z = X \cup Y$. Prove that $\overline{Z} = \overline{X} \cup \overline{Y}$.

Proof: Let X, Y, and Z be sets of real numbers with $Z = X \cup Y$. Let $x \in \overline{Z}$. By part 3 of Theorem 7.25, $x \in Z$ or x is an accumulation point of Z. If $x \in Z$, then $x \in X$ or $x \in Y$. By part 1 of Theorem 7.25, $X \subseteq \overline{X}$ and $Y \subseteq \overline{Y}$. So, $x \in \overline{X}$ or $x \in \overline{Y}$. Thus, $x \in \overline{X} \cup \overline{Y}$. Now, suppose that x is an accumulation point of Z. Without loss of generality, assume that x is **not** an accumulation point of Y. Then there is an open interval (c, d) with $x \in (c, d)$ and such that if $y \in Y$ with $y \neq x$, then $y \notin (c, d)$. Let (a, b) be any open interval containing x and let $U = (a, b) \cap (c, d)$. Since $x \in U$, U is a nonempty open interval. Since x is an accumulation point of Z, there is $y \in Z$ with $y \in U$ and $y \neq x$. Since $U \subseteq (c, d)$, $y \notin Y$. Therefore, $y \in X$. Since $U \subseteq (a, b)$, we have $y \in (a, b)$. It follows that x is an accumulation point of X. So, $x \in \overline{X}$. Therefore, $x \in \overline{X}$ or $x \in \overline{Y}$, and it follows that $x \in \overline{X} \cup \overline{Y}$. Since $x \in \overline{Z}$ was arbitrary, $\overline{Z} \subseteq \overline{X} \cup \overline{Y}$.

Conversely, let $x \in \overline{X} \cup \overline{Y}$. Then $x \in \overline{X}$ or $x \in \overline{Y}$. Without loss of generality, assume that $x \in \overline{X}$. By part 3 of Theorem 7.25, $x \in X$ or x is an accumulation point of X. If $x \in X$, then $x \in X$ or $x \in Y$. So, $x \in X \cup Y = Z$. By part 1 of Theorem 7.25, $Z \subseteq \overline{Z}$. So, $x \in \overline{Z}$. Now, suppose that x is an accumulation point of X and let (a, b) be any open interval containing x. Then there is $y \in X$ with $y \in (a, b)$ and $y \neq x$. So, $y \in X$ or $y \in Y$, and thus, $y \in X \cup Y = Z$. It follows that x is an accumulation point of Z. Therefore, $x \in \overline{Z}$. Since $x \in \overline{X} \cup \overline{Y}$ was arbitrary, $\overline{X} \cup \overline{Y} \subseteq \overline{Z}$.

Since $\overline{Z} \subseteq \overline{X} \cup \overline{Y}$ and $\overline{X} \cup \overline{Y} \subseteq \overline{Z}$, we have $\overline{Z} = \overline{X} \cup \overline{Y}$. $\square$

13. Prove that if X is a nonempty set of closed subsets of $\mathbb{R}$, then $\cap X$ is closed.

Proof: Let X be a nonempty set of closed subsets of $\mathbb{R}$. Then for each $A \in X$, $\mathbb{R} \setminus A$ is an open set in $\mathbb{R}$. By Theorem 7.15, $\cup \{\mathbb{R} \setminus A \mid A \in X\}$ is open in $\mathbb{R}$. Therefore, $\mathbb{R} \setminus \cup \{\mathbb{R} \setminus A \mid A \in X\}$ is closed in $\mathbb{R}$. So, it suffices to show that $\cap X = \mathbb{R} \setminus \cup \{\mathbb{R} \setminus A \mid A \in X\}$. Well, $x \in \cap X$ if and only if for all $A \in X$, $x \in A$ if and only if for all $A \in X$, $x \notin \mathbb{R} \setminus A$ if and only if $x \notin \cup \{\mathbb{R} \setminus A \mid A \in X\}$ if and only if $x \in \mathbb{R} \setminus \cup \{\mathbb{R} \setminus A \mid A \in X\}$. So, $\cap X = \mathbb{R} \setminus \cup \{\mathbb{R} \setminus A \mid A \in X\}$, completing the proof. $\square$

14. Prove that a closed subset of a compact set is compact.

Proof: Let $A \subseteq B \subseteq \mathbb{R}$ with A closed and B compact. Let $\mathcal{C}$ be an open covering of A. Then we have that $\mathcal{D} = \mathcal{C} \cup \{\mathbb{R} \setminus A\}$ is an open covering of B. By the compactness of B, there is a finite subcollection $\mathcal{E}$ of $\mathcal{D}$ such that $\mathcal{E}$ is an open covering of B. It follows that $\mathcal{E} \setminus \{\mathbb{R} \setminus A\}$ is an open covering of A. So, A is compact. $\square$

15. Give an example of an infinite collection of open sets in $\mathbb{R}$ whose intersection is not open in $\mathbb{R}$. Also, give an example of an infinite collection of closed sets in $\mathbb{R}$ whose union is not closed in $\mathbb{R}$. Provide a proof for each example.

Solution: Let $X = \left\{ \left(0, 1 + \frac{1}{n}\right) \mid n \in \mathbb{Z}^+ \right\}$. Each set in X is an open interval, and therefore, open. We will show that $\cap X = (0, 1]$. Note that $x \in \cap X$ if and only if for all $n \in \mathbb{Z}^+$, $x \in \left(0, 1 + \frac{1}{n}\right)$ if and only if for all $n \in \mathbb{Z}^+$, $0 < x < 1 + \frac{1}{n}$. We need to show that $x \leq 1$ is equivalent to $\forall n \in \mathbb{Z}^+ \left(x < 1 + \frac{1}{n}\right)$.

Suppose that $x \leq 1$. Let $n \in \mathbb{Z}^+$. By Theorem 6.8, $\frac{1}{n} > 0$. So, $1 + \frac{1}{n} - 1 > 0$ (SACT). Thus, $1 + \frac{1}{n} > 1$. So, we have $x \leq 1 < 1 + \frac{1}{n}$, and therefore, $x < 1 + \frac{1}{n}$. Since $n \in \mathbb{Z}^+$ was arbitrary, we have shown that $x \leq 1$ implies $\forall n \in \mathbb{Z}^+ \left(x < 1 + \frac{1}{n}\right)$.

Now, suppose $x > 1$ (proof by contrapositive). Then $x - 1 > 0$. Since there is no smallest positive real number, there is a real number $\epsilon > 0$ with $x - 1 > \epsilon$. By the Archimedean Property of the reals, there is a natural number n with $n > \frac{1}{\epsilon}$. So, $\frac{1}{n} < \epsilon$, or equivalently, $\epsilon > \frac{1}{n}$. Thus, $x - 1 > \frac{1}{n}$, and so, $x > 1 + \frac{1}{n}$. We have shown that there is $n \in \mathbb{Z}^+$ such that $x \geq 1 + \frac{1}{n}$. So, $\forall n \in \mathbb{Z}^+ \left(x < 1 + \frac{1}{n}\right)$ is false.

This equivalence proves that $\cap X = (0, 1]$, an interval that is **not** an open set.

Let $Y = \left\{ \left[0, 1 - \frac{1}{n}\right] \mid n \in \mathbb{Z}^+ \right\}$. Each set in Y is a closed interval, and therefore, closed. We will show that $\cup Y = [0, 1)$. Note that $x \in \cup Y$ if and only if there is $n \in \mathbb{Z}^+$ such that $x \in \left[0, 1 - \frac{1}{n}\right]$ if and only if there is $n \in \mathbb{Z}^+$ such that $0 \leq x \leq 1 - \frac{1}{n}$. We need to show that $x < 1$ is equivalent to $\exists n \in \mathbb{Z}^+ \left(x \leq 1 - \frac{1}{n}\right)$ (where $\exists$ is read "there exists" or "there is").

Suppose $x < 1$. Then $1 - x > 0$. Since there is no smallest positive real number, there is a real number $\epsilon > 0$ with $1 - x > \epsilon$. By the Archimedean Property of the reals, there is a natural number n with $n > \frac{1}{\epsilon}$. So, $\frac{1}{n} < \epsilon$, or equivalently, $\epsilon > \frac{1}{n}$. Thus, $1 - x > \frac{1}{n}$, and so, $x < 1 - \frac{1}{n}$. We have shown that there is $n \in \mathbb{Z}^+$ such that $x \leq 1 - \frac{1}{n}$. So, $\exists n \in \mathbb{Z}^+ \left(x \leq 1 - \frac{1}{n}\right)$.

Now, suppose $x \geq 1$ (proof by contrapositive). Let $n \in \mathbb{Z}^+$. By Theorem 6.8, $\frac{1}{n} > 0$. So, $1 - 1 + \frac{1}{n} > 0$ (SACT). So, $1 > 1 - \frac{1}{n}$. It follows that $x > 1 - \frac{1}{n}$. Since $n \in \mathbb{Z}^+$ was arbitrary, $\forall n \in \mathbb{Z}^+ \left(x > 1 - \frac{1}{n}\right)$. It follows that $\exists n \in \mathbb{Z}^+ \left(x \leq 1 - \frac{1}{n}\right)$ is false.

This equivalence proves that $\cup Y = [0, 1)$, an interval that is **not** a closed set.

16. Determine if each of the following subsets of $\mathbb{R}$ is open, closed, both, or neither. Give a proof in each case.

   (i) $\{x \in \mathbb{R} \mid |x| > 1\}$

   (ii) $\{(-1)^n \mid n \in \mathbb{Z}^+\}$

   (iii) $\{x \in \mathbb{R} \mid 2 < |x - 2| < 4\}$

Proofs:

(i) $S = \{x \in \mathbb{R} \mid |x| > 1\}$ is **open**. To see this, let $y \in S$ and let $r = |y| - 1$. We will show that $N_r(y) \subseteq S$ (recall that $N_r(y) = \{x \in \mathbb{R} \mid |x - y| < r\}$). Let $x \in N_r(y)$. Then we have $|x - y| < r = |y| - 1$. So, using the Triangle Inequality, we have

$$|y| = |(y - x) + x| \le |y - x| + |x| = |x - y| + |x| < |y| - 1 + |x|.$$

Thus, $|y| < |y| - 1 + |x|$, and therefore, $|x| > 1$. So, $x \in S$. Since $x \in N_r(y)$ was arbitrary, we have shown that $N_r(y) \subseteq S$. So, S is open. $\square$

$S = \{x \in \mathbb{R} \mid |x| > 1\}$ is **not closed** because $\mathbb{R} \setminus S = \{x \in \mathbb{R} \mid |x| \le 1\}$ is not open.

(ii) Note that $S = \{(-1)^n \mid n \in \mathbb{Z}^+\}$ is a finite set consisting of just two real numbers. Indeed, $S = \{1, -1\}$.

S is **not open**. To see this, let $N_r(1)$ be an arbitrary r-neighborhood of 1. Then $1 + \frac{r}{2} \in N_r(1)$ because $\left|\left(1 + \frac{r}{2}\right) - 1\right| = \left|\frac{r}{2}\right| = \frac{r}{2} < r$, but $1 + \frac{r}{2} \notin S$ because $1 + \frac{r}{2}$ is not equal to 1 or -1.

S is **closed**. To see this, we show that $T = \mathbb{R} \setminus S$ is open.

Let $y \in T$ and let $r = \min\{|y - 1|, |y + 1|\}$.

We now show that $N_r(y) \subseteq T$. Since $r \le |y - 1|$, $1 \notin N_r(y)$ (otherwise, $|y - 1| < r$). Similarly, $-1 \notin N_r(y)$. So, if $x \in N_r(y)$, then $x \notin S$, and so, $x \in T$.

Thus, $T = \mathbb{R} \setminus S$ is open. Therefore, S is closed. $\square$

(iii) $S = \{x \in \mathbb{R} \mid 2 < |x - 2| < 4\}$ is **open** and **not closed**.

To see that S is open, let $x \in S$ and let $r = \min\{|x - 2| - 2, 4 - |x - 2|\}$. We show that $N_r(x) \subseteq S$. Let $y \in N_r(x)$. Then $|y - x| < r$. So, $|y - x| < 4 - |x - 2|$. Therefore, we have $|y - x| + |x - 2| < 4$, and so, $|y - 2| = |y - x + x - 2| \le |y - x| + |x - 2| < 4$.

Also, $|y - x| < |x - 2| - 2$. So, we have

$$2 < |x - 2| - |y - x| = |x - y + y - 2| - |y - x| \le |x - y| + |y - 2| - |y - x|$$
$$= |y - x| + |y - 2| - |y - x| = |y - 2|.$$

So, $2 < |y - 2| < 4$, and therefore, $y \in S$.

Since $x \in S$ was arbitrary, S is open.

To see that S is not closed, we show that $\mathbb{R} \setminus S = \{x \in \mathbb{R} \mid |x - 2| \leq 2 \text{ or } |x - 2| \geq 4\}$ is not open. To see this, first note that $|6 - 2| = |4| = 4 \geq 4$, and so, $6 \in \mathbb{R} \setminus S$. Let $N_r(6)$ be an r-neighborhood of 6 and let $k = \min\left\{1, \frac{r}{2}\right\}$. Then we have $6 - k \in N_r(6)$ because $|(6 - k) - 6| = |-k| = k \leq \frac{r}{2} < r$. However, $6 - k \notin \mathbb{R} \setminus S$. To see this, first observe that $|(6 - k) - 2| = |4 - k|$. If $k = 1$, then $|4 - k| = |4 - 1| = |3| = 3$ and it follows that $6 - k \notin \mathbb{R} \setminus S$. If $k = \frac{r}{2}$, then $0 < \frac{r}{2} \leq 1$, so that $-1 \leq -\frac{r}{2} < 0$, and thus, $3 < 4 - \frac{r}{2} < 4$. So, $3 < \left|4 - \frac{r}{2}\right| < 4$ and once again, $6 - k \notin \mathbb{R} \setminus S$. So, $\mathbb{R} \setminus S$ is not open. Therefore, S is not closed. $\quad\square$

LEVEL 5

17. Prove that every closed set in $\mathbb{R}$ can be written as an intersection $\cap X$, where each element of X is a union of at most 2 closed intervals.

Proof: First note that $\mathbb{R} = \cap\{\mathbb{R}\}$.

Let A be a closed set in $\mathbb{R}$ with $A \neq \mathbb{R}$. Then $\mathbb{R} \setminus A$ is a nonempty open set in $\mathbb{R}$. By Theorem 7.17, $\mathbb{R} \setminus A$ can be expressed as $\cup X$, where X is a set of bounded open intervals. For each B in X, $\mathbb{R} \setminus B$ is a union of two closed intervals (if $B = (a, b)$, then $\mathbb{R} \setminus B = (-\infty, a] \cup [b, \infty)$). Now, by part (iii) of Problem 27 from Problem Set 1, we have $A = \mathbb{R} \setminus (\mathbb{R} \setminus A) = \mathbb{R} \setminus \cup X = \cap\{\mathbb{R} \setminus B \mid B \in X\}$. $\quad\square$

18. A real number x is an **interior point** of a set S of real numbers if there is a neighborhood of x that contains only points in S, whereas x is a **boundary point** of S if each neighborhood of x contains at least one point in S and one point not in S. Prove the following:

   (i)   A set of real numbers is open in $\mathbb{R}$ if and only if each point in the set is an interior point of the set.

   (ii)   A set of real numbers is open in $\mathbb{R}$ if and only if it contains none of its boundary points.

   (iii) A set of real numbers is closed in $\mathbb{R}$ if and only if it contains all its boundary points.

Proofs:

   (i)   Let S be a set of real numbers. Then S is open if and only if for every real number $x \in S$, there is an open interval (a, b) with $x \in (a, b)$ and $(a, b) \subseteq S$ if and only if for every real number $x \in S$, there is a neighborhood of x that contains only points in S if and only if every real number in S is an interior point of S. $\quad\square$

   (ii)   Suppose that S is an open set of real numbers and let $x \in S$. By (i), x is an interior point of S. So, there is a neighborhood of x containing only points of S. So, x is **not** a boundary point of S. Since $x \in S$ was arbitrary, S contains none of its boundary points.

   We now prove that if S contains none of its boundary points, then S is open by contrapositive. Suppose S is not open. By (i), there is $x \in S$ such that x is **not** an interior point. Let N be a neighborhood of x. Since $x \in S$, N contains a point in S (namely, x). Since x is not an interior point of S, N contains a point not in S. So, x is a boundary point of S. Therefore, S contains at least one of its boundary points. $\quad\square$

(iii) First note that a real number x is a boundary point of S if and only if x is a boundary point of $\mathbb{R} \setminus S$ (because $x \in S$ if and only if $x \notin \mathbb{R} \setminus S$, and vice versa).

Let S be a set of real numbers. Then S is closed if and only if $\mathbb{R} \setminus S$ is open if and only if $\mathbb{R} \setminus S$ contains none of its boundary points (by (ii)) if and only if $S = \mathbb{R} \setminus (\mathbb{R} \setminus S)$ contains all its boundary points. □

19. Let C be the Cantor set. Prove the following:

(i) $\frac{1}{4} \in C$.

(ii) C is uncountable.

Proofs:

(i) Recall that $C = \cap \{C_n \mid n \in \mathbb{N}\}$. First note that if $x \in C_n$, then $\frac{1}{3}x \in C_{n+1}$. Also, note that $x \in C_n$ if and only if $1 - x \in C_n$. Now, we prove by induction that for all $n \in \mathbb{N}$, $\frac{1}{4} \in C_n$ and $\frac{3}{4} \in C_n$.

Base case ($k = 0$): Clearly $\frac{1}{4}, \frac{3}{4} \in C_0 = [0, 1]$.

Inductive step: Assume that $\frac{1}{4}, \frac{3}{4} \in C_k$. Since $\frac{3}{4} \in C_k$, $\frac{1}{4} = \frac{1}{3} \cdot \frac{3}{4} \in C_{k+1}$. By the symmetry of C_{k+1}, $\frac{3}{4} = 1 - \frac{1}{4} \in C_{k+1}$.

By the Principle of Mathematical Induction, for all $n \in \mathbb{N}$, $\frac{1}{4} \in C_n$ and $\frac{3}{4} \in C_n$.

Therefore, $\frac{1}{4}, \frac{3}{4} \in \cap \{C_n \mid n \in \mathbb{N}\} = C$. □

(ii) We define an injection $f: [0, 1) \to C$ as follows. For $x \in [0, 1)$, let $x = 0.x_0x_1x_2 \ldots$ be the binary expansion of x (see Note 2 following Example 5.7). For each $n \in \mathbb{N}$, let $z_n = 2x_n$ and consider $z = 0.z_0z_1z_2 \ldots$ as the ternary expansion of a real number. Observe that for all $n \in \mathbb{N}$, $z_n = 0$ or $z_n = 2$. Since $z_0 \neq 1$, $z \notin \left[\frac{1}{3}, \frac{2}{3}\right)$. So $z \in C_1$. Since $z_1 \neq 1$, $z \notin \left[\frac{1}{9}, \frac{2}{9}\right)$ and $z \notin \left[\frac{7}{9}, \frac{8}{9}\right)$. So, $z \in C_2$. By a continuation of this reasoning, we see that for all $n \in \mathbb{N}$, $z \in C_n$. Therefore, $z \in \cap\{C_n \mid n \in \mathbb{N}\} = C$. So, if we let $f(x) = z$, then $f: [0, 1) \to C$. To see that f is injective, observe that if $x \neq y$, then for some $n \in \mathbb{N}$, $x_n \neq y_n$. Without loss of generality, assume that $x_n = 0$ and $y_n = 1$. So, if $f(x) = z$ and $f(y) = w$, then $z_n = 0$ and $w_n = 2$. It follows that $f(x) = y \neq w = f(y)$. □

Problem Set 8

LEVEL 1

1. Let $f: \mathbb{R} \to \mathbb{R}$ be defined by $f(x) = 5x - 1$.

    (i) Prove that $\lim\limits_{x \to 3} f(x) = 14$.

    (ii) Prove that f is continuous on $\mathbb{R}$.

Proofs:

(i) Let $\epsilon > 0$ and let $\delta = \frac{\epsilon}{5}$. Suppose that $0 < |x - 3| < \delta$. Then we have

$$|(5x - 1) - 14| = |5x - 15| = 5|x - 3| < 5\delta = 5 \cdot \frac{\epsilon}{5} = \epsilon.$$

So, $\lim\limits_{x \to 3} f(x) = 14$. □

(ii) Let $a \in \mathbb{R}$. We will show that f is continuous at a. Let $\epsilon > 0$, let $\delta = \frac{\epsilon}{5}$, and let $|x - a| < \delta$. Then

$$|(5x - 1) - (5a - 1)| = |5x - 5a| = 5|x - a| < 5\delta = 5 \cdot \frac{\epsilon}{5} = \epsilon.$$

So, f is continuous at a. Since $a \in \mathbb{R}$ was arbitrary, f is continuous on $\mathbb{R}$. □

2. Let $r, c \in \mathbb{R}$ and let $f: \mathbb{R} \to \mathbb{R}$ be defined by $f(x) = c$. Prove that $\lim\limits_{x \to r}[f(x)] = c$.

Proof: Let $\epsilon > 0$ and let $\delta = 1$. If $0 < |x - r| < \delta$, then $|f(x) - c| = |c - c| = |0| = 0 < \epsilon$. Therefore, $\lim\limits_{x \to r}[f(x)] = c$. □

3. Let $A \subseteq \mathbb{R}$, let $f: A \to \mathbb{R}$, let $r, k \in \mathbb{R}$, and suppose that $\lim\limits_{x \to r}[f(x)]$ is a finite real number. Prove that $\lim\limits_{x \to r}[kf(x)] = k \lim\limits_{x \to r}[f(x)]$.

Proof: Suppose that $\lim\limits_{x \to r}[f(x)] = L$ and let $\epsilon > 0$. First assume that $k \neq 0$. Since $\lim\limits_{x \to r}[f(x)] = L$, there is $\delta > 0$ such that $0 < |x - r| < \delta$ implies $|f(x) - L| < \frac{\epsilon}{|k|}$. Suppose that $0 < |x - r| < \delta$. Then

$$|kf(x) - kL| = |k||f(x) - L| < |k|\frac{\epsilon}{|k|} = \epsilon.$$

So, $\lim\limits_{x \to r}[kf(x)] = kL = k \lim\limits_{x \to r}[f(x)]$.

If $k = 0$, let $\delta = 1$. If $0 < |x - r| < \delta$, then

$$|kf(x) - kL| = |0f(x) - 0L| = |0| = 0 < \epsilon.$$

So, in this case, we also have $\lim\limits_{x \to r}[kf(x)] = kL = k \lim\limits_{x \to r}[f(x)]$. □

4. Prove that a real-valued sequence (s_n) is bounded if and only if there is $M \in \mathbb{R}^+$ such that for all $m, n \in \mathbb{N}$, $|s_m - s_n| \leq M$.

Proof: Suppose that (s_n) is bounded in $\mathbb{R}$. Then there is $K \in \mathbb{R}^+$ such that $|s_n| \leq K$ for all $n \in \mathbb{N}$. So, for all $m, n \in \mathbb{N}$, we have $|s_m - s_n| \leq |s_m| + |s_n| \leq 2K$ (by the Triangle Inequality). Let $M = 2K$. Then $M \in \mathbb{R}^+$ and for all $m, n \in \mathbb{N}$, we have $|s_m - s_n| \leq M$.

Conversely, suppose that there is $L \in \mathbb{R}^+$ such that for all $m, n \in \mathbb{N}$, $|s_m - s_n| \leq L$. By the Triangle Inequality (and SACT), we have

$$|s_m| = |(s_m - s_0) + s_0| \leq |s_m - s_0| + |s_0| \leq L + |s_0|.$$

So, if we let $M = L + |s_0|$. Then for all $m \in \mathbb{N}$, $|s_m| \leq M$. So, (s_n) is bounded. $\square$

LEVEL 2

5. Let $A \subseteq \mathbb{R}$, let $f: A \to \mathbb{R}$, and let r be an interior point of A (see Problem 18 from Problem Set 7). Prove that f is continuous at r if and only if $\lim\limits_{x \to r}[f(x)] = f(r)$.

Proof: Let $A \subseteq \mathbb{R}$, let $f: A \to \mathbb{R}$, and let r be an interior point of A. First suppose that f is continuous at r. Let $\epsilon > 0$. Then there is $\delta > 0$ such that $|x - r| < \delta$ implies $|f(x) - f(r)| < \epsilon$. Let $x \in \mathbb{R}$ satisfy $0 < |x - r| < \delta$. Then $|x - r| < \delta$. So, $|f(x) - f(r)| < \epsilon$. Since $\epsilon > 0$ was arbitrary, $\lim\limits_{x \to r}[f(x)] = f(r)$.

Now, suppose that $\lim\limits_{x \to r}[f(x)] = f(r)$. Let $\epsilon > 0$. Then there is $\delta > 0$ such that $0 < |x - r| < \delta$ implies $|f(x) - f(r)| < \epsilon$. Let $x \in \mathbb{R}$ satisfy $|x - r| < \delta$. Then $0 < |x - r| < \delta$ or $x = r$. If $0 < |x - r| < \delta$, then $|f(x) - f(r)| < \epsilon$. If $x = r$, then $|f(x) - f(r)| = |f(r) - f(r)| = |0| = 0 < \epsilon$. Since $\epsilon > 0$ was arbitrary, f is continuous at r. $\square$

6. Prove that every polynomial function $p: \mathbb{R} \to \mathbb{R}$ is continuous on $\mathbb{R}$.

Proof: Let $r \in \mathbb{R}$. We first show by induction that for all $n \in \mathbb{N}$ with $n \geq 1$, $\lim\limits_{x \to r}[x^n] = r^n$.

Base case $(k = 1)$: Let $\epsilon > 0$ be given and let $\delta = \epsilon$. Then $0 < |x - r| < \delta$ implies $|x - r| < \delta = \epsilon$. Since $\epsilon > 0$ was arbitrary, $\lim\limits_{x \to r}[x] = r$.

Inductive step: Let $k \in \mathbb{N}$ and assume that $\lim\limits_{x \to r}[x^k] = r^k$. By Theorem 8.14, we have

$$\lim_{x \to r}[x^{k+1}] = \lim_{x \to r}[x^k \cdot x] = \lim_{x \to r}[x^k] \cdot \lim_{x \to r}[x] = r^k \cdot r = r^{k+1}.$$

By the Principle of Mathematical Induction, for all $n \in \mathbb{N}$ with $n \geq 1$, $\lim\limits_{x \to r}[x^n] = r^n$.

Now, let $p: \mathbb{R} \to \mathbb{R}$ be a polynomial, say $p(x) = a_n x^n + a_{n-1} x^{n-1} + \cdots + a_1 x + a_0$. By Problem 2, $\lim\limits_{x \to r}[a_0] = a_0$. By the last paragraph and Problem 3, $\lim\limits_{x \to r}[a_k x^k] = a_k \lim\limits_{x \to r}[x^k] = a_k r^k$. Finally, using Theorem 8.13, we have

$$\lim_{x \to r}[p(x)] = \lim_{x \to r}[a_n x^n + a_{n-1} x^{n-1} + \cdots + a_1 x + a_0]$$
$$= \lim_{x \to r}[a_n x^n] + \lim_{x \to r}[a_{n-1} x^{n-1}] + \cdots + \lim_{x \to r}[a_1 x] + \lim_{x \to r}[a_0]$$
$$= a_n r^n + a_{n-1} r^{n-1} + \cdots + a_1 r + a_0 = p(r).$$

By Problem 5, p is continuous at r. Since $r \in \mathbb{R}$ was arbitrary, p is continuous on $\mathbb{R}$. $\quad\square$

LEVEL 3

7. Let $g \colon \mathbb{R} \to \mathbb{R}$ be defined by $g(x) = 2x^2 - 3x + 7$.

   (i) Prove that $\lim_{x \to 1} g(x) = 6$.

   (ii) Prove that g is continuous on $\mathbb{R}$.

Proofs:

(i) Let $\epsilon > 0$ and let $\delta = \min\left\{1, \frac{\epsilon}{3}\right\}$. Suppose that $0 < |x - 1| < \delta$. Then we have $|x - 1| < 1$, so that $-1 < x - 1 < 1$. Adding 1, we get $0 < x < 2$. Multiplying by 2, we have $0 < 2x < 4$. Subtracting 1 gives us $-1 < 2x - 1 < 3$. So, $-3 < 2x - 1 < 3$, and therefore, $|2x - 1| < 3$. Now, we have

$$|(2x^2 - 3x + 7) - 6| = |2x^2 - 3x + 1| = |2x - 1||x - 1| < 3\delta \le 3 \cdot \frac{\epsilon}{3} = \epsilon.$$

So, $\lim_{x \to 1} g(x) = 6$. $\quad\square$

(ii) Let $a \in \mathbb{R}$. We will show that f is continuous at a. Let $\epsilon > 0$ and let $\delta = \min\left\{1, \frac{\epsilon}{M}\right\}$, where $M = \max\{|4a - 8|, |4a - 4|\}$. Suppose that $|x - a| < \delta$. Then we have $|x - a| < 1$, so that $-1 < x - a < 1$. Adding $2a - 3$, we get $2a - 4 < x + a - 3 < 2a - 2$. Multiplying by 2 yields $4a - 8 < 2(x + a - 3) < 4a - 4$. Therefore, $-M < 2(x + a - 3) < M$, or equivalently, $|2(x + a - 3)| < M$. Now, we have

$$|(2x^2 - 3x + 7) - (2a^2 - 3a + 7)| = |2(x^2 - a^2) - 3(x - a)| = |x - a||2(x + a - 3)|$$
$$< \delta M \le \frac{\epsilon}{M} \cdot M = \epsilon.$$

So, g is continuous at a. Since $a \in \mathbb{R}$ was arbitrary, g is continuous on $\mathbb{R}$. $\quad\square$

8. Suppose that $f, g \colon \mathbb{R} \to \mathbb{R}$, $a \in \mathbb{R}$, f is continuous at a, and g is continuous at $f(a)$. Prove that $g \circ f$ is continuous at a.

Proof: Let $f, g \colon \mathbb{R} \to \mathbb{R}$, let $a \in \mathbb{R}$, and suppose that f is continuous at a and g is continuous at $f(a)$. Let $\epsilon > 0$. Since g is continuous at $f(a)$, there is $\delta_1 > 0$ such that $|y - f(a)| < \delta_1$ implies $|g(y) - g(f(a))| < \epsilon$. Since f is continuous at a, there is $\delta_2 > 0$ such that $|x - a| < \delta_2$ implies $|f(x) - f(a)| < \delta_1$. Now, suppose that $|x - a| < \delta_2$. Then $|f(x) - f(a)| < \delta_1$. It follows that $|g(f(x)) - g(f(a))| < \epsilon$. Since $\epsilon > 0$ was arbitrary, $g \circ f$ is continuous at a. $\quad\square$

9. Let $h: \mathbb{R} \to \mathbb{R}$ be defined by $h(x) = \frac{x^3-4}{x^2+1}$. Prove that $\lim_{x \to 2} h(x) = \frac{4}{5}$.

Proof: Let $\epsilon > 0$ and let $\delta = \min\left\{1, \frac{2\epsilon}{15}\right\}$. Suppose that $0 < |x - 2| < \delta$. Then we have $|x - 2| < 1$, so that $-1 < x - 2 < 1$. Adding 2, we get $1 < x < 3$. So, $23 < 5x^2 + 6x + 12 < 75$ and therefore, $-75 < 5x^2 + 6x + 12 < 75$. So, $|5x^2 + 6x + 12| < 75$. Also, $2 < x^2 + 1 < 10$. In particular, we have $x^2 + 1 > 2$, and so, $\frac{1}{x^2+1} < \frac{1}{2}$. Now, we have

$$\left|\frac{x^3-4}{x^2+1} - \frac{4}{5}\right| = \left|\frac{5(x^3-4)}{5(x^2+1)} - \frac{4(x^2+1)}{5(x^2+1)}\right| = \left|\frac{5x^3 - 4x^2 - 24}{5(x^2+1)}\right|$$

$$= \frac{|5x^2 + 6x + 12||x - 2|}{5(x^2+1)} < \frac{75\delta}{5 \cdot 2} \le \frac{75}{10} \cdot \frac{2\epsilon}{15} = \epsilon.$$

So, $\lim_{x \to 2} h(x) = \frac{4}{5}$. $\qquad\qquad\square$

10. Let $k: (0, \infty) \to \mathbb{R}$ be defined by $k(x) = \sqrt{x}$.

   (i)   Prove that $\lim_{x \to 25} k(x) = 5$.

   (ii) Prove that f is continuous on $(0, \infty)$.

   (iii) Is f uniformly continuous on $(0, \infty)$?

Proofs:

   (i) Let $\epsilon > 0$ and let $\delta = \min\{1, (5 + \sqrt{24})\epsilon\}$. Suppose that $0 < |x - 25| < \delta$. Then we have $|x - 25| < 1$, so that $-1 < x - 25 < 1$. Adding 25, we get $24 < x < 26$. Taking square roots, we have $\sqrt{24} < \sqrt{x} < \sqrt{26}$. Adding 5 gives us $5 + \sqrt{24} < \sqrt{x} + 5 < 5 + \sqrt{26}$. So, $\frac{1}{\sqrt{x}+5} < \frac{1}{5+\sqrt{24}}$. Now, we have

$$\left|\sqrt{x} - 5\right| = \left|\frac{(\sqrt{x} - 5)(\sqrt{x} + 5)}{\sqrt{x} + 5}\right| = \frac{|x - 25|}{\sqrt{x} + 5} < \frac{\delta}{5 + \sqrt{24}} \le \frac{1}{5 + \sqrt{24}} \cdot (5 + \sqrt{24})\epsilon = \epsilon.$$

So, $\lim_{x \to 25} k(x) = 5$. $\qquad\qquad\square$

   (ii) Let $a \in \mathbb{R}$. We will show that f is continuous at a. Let $\epsilon > 0$, let $\delta = \min\{1, \epsilon\sqrt{a}\}$, and let $x \in (0, \infty)$ satisfy $|x - a| < \delta$. Then we have $|x - a| < 1$, so that $-1 < x - a < 1$. Adding a, we get $a - 1 < x < a + 1$. Since $x \in (0, \infty)$, we have $0 < x < a + 1$. Taking square roots, we have $0 < \sqrt{x} < \sqrt{a+1}$. Adding $\sqrt{a}$ gives us $\sqrt{a} < \sqrt{x} + \sqrt{a} < \sqrt{a+1} + \sqrt{a}$. Therefore, $\frac{1}{\sqrt{x}+\sqrt{a}} < \frac{1}{\sqrt{a}}$. Now, we have

$$\left|\sqrt{x} - \sqrt{a}\right| = \left|\frac{(\sqrt{x} - \sqrt{a})(\sqrt{x} + \sqrt{a})}{\sqrt{x} + \sqrt{a}}\right| = \frac{|x - a|}{\sqrt{x} + \sqrt{a}} < \frac{\delta}{\sqrt{a}} \le \frac{\epsilon\sqrt{a}}{\sqrt{a}} = \epsilon.$$

So, f is continuous at a. Since $a \in \mathbb{R}$ was arbitrary, f is continuous on $\mathbb{R}$. $\qquad\square$

(iii) Let $\epsilon > 0$, let $\delta = \epsilon^2$, and let $x, y \in (0, \infty)$ satisfy $|x - y| < \delta$. Then we have

$$|\sqrt{x} - \sqrt{y}| = \sqrt{(\sqrt{x} - \sqrt{y})^2} \le \sqrt{|\sqrt{x} - \sqrt{y}||\sqrt{x} + \sqrt{y}|} = \sqrt{|x - y|} < \sqrt{\delta} = \sqrt{\epsilon^2} = \epsilon.$$

So, f is uniformly continuous on $\mathbb{R}$. $\quad\square$

11. Let $f: \mathbb{R} \to \mathbb{R}$ be defined by $f(x) = x^2$. Prove that f is continuous on $\mathbb{R}$, but not uniformly continuous on $\mathbb{R}$.

Proof: Let $a \in \mathbb{R}$. We will show that f is continuous at a. Let $\epsilon > 0$ and let $\delta = \min\left\{1, \frac{\epsilon}{M}\right\}$, where $M = \max\{|2a - 1|, |2a + 1|\}$. Suppose that $|x - a| < \delta$. Then $|x - a| < 1$, so that $-1 < x - a < 1$. Adding $2a$, we get $2a - 1 < x + a < 2a + 1$. So, $-M < x + a < M$, or equivalently, $|x + a| < M$. Now, we have

$$|x^2 - a^2| = |x - a||x + a| < \delta \cdot M \le \frac{\epsilon}{M} \cdot M = \epsilon.$$

So, f is continuous at a. Since $a \in \mathbb{R}$ was arbitrary, f is continuous on $\mathbb{R}$.

To see that f is not uniformly continuous, let $\epsilon = 1$ and let $\delta > 0$. Let $x = \frac{1}{\delta}$ and $y = \frac{1}{\delta} + \frac{\delta}{2}$. Then we have $|x - y| = \left|\frac{1}{\delta} - \left(\frac{1}{\delta} + \frac{\delta}{2}\right)\right| = \frac{\delta}{2}$, but

$$|f(x) - f(y)| = |x^2 - y^2| = \left|\frac{1}{\delta^2} - \left(\frac{1}{\delta} + \frac{\delta}{2}\right)^2\right| = \left|\frac{1}{\delta^2} - \frac{1}{\delta^2} - 1 - \frac{\delta^2}{4}\right| = 1 + \frac{\delta^2}{4} > 1 = \epsilon.$$

So, f is **not** uniformly continuous on $\mathbb{R}$ (and in fact, not uniformly continuous on $(0, \infty)$ since we only needed positive values of x and y to violate the definition of uniform continuity). $\quad\square$

12. Let $A \subseteq \mathbb{R}$, let $f: A \to \mathbb{R}$, let $r \in \mathbb{R}$, and suppose that $\lim_{x \to r}[f(x)] > 0$. Prove that there is a deleted neighborhood N of r such that $f(x) > 0$ for all $x \in N$.

Proof: Suppose that $\lim_{x \to r}[f(x)] = L$ with $L > 0$. Let $\epsilon = \frac{L}{2}$. There is $\delta > 0$ such that $0 < |x - r| < \delta$ implies $|f(x) - L| < \epsilon$. Consider $N_\delta^\odot(r) = (r - \delta, r) \cup (r, r + \delta)$. Let $x \in N_\delta^\odot(r)$. Then we have $x \in (r - \delta, r) \cup (r, r + \delta)$, so that $0 < |x - r| < \delta$. It follows that $|f(x) - L| < \epsilon = \frac{L}{2}$. So, we have $-\frac{L}{2} < f(x) - L < \frac{L}{2}$, or equivalently, $L - \frac{L}{2} < f(x) < L + \frac{L}{2}$. Since $L - \frac{L}{2} = \frac{L}{2}$ and $L + \frac{L}{2} = \frac{3L}{2}$, we have $\frac{L}{2} < f(x) < \frac{3L}{2}$. In particular, we have $f(x) > \frac{L}{2} > 0$. Since $x \in N_\delta^\odot(r)$ was arbitrary, we have shown that for all $x \in N_\delta^\odot(r)$, $f(x) > 0$. $\quad\square$

13. Let $A \subseteq \mathbb{R}$, let $f: A \to \mathbb{R}$, let $r \in \mathbb{R}$, and suppose that $\lim_{x \to r}[f(x)]$ is a finite real number. Prove that there is $M \in \mathbb{R}$ and an open interval (a, b) containing r such that $|f(x)| \le M$ for all $x \in (a, b) \setminus \{r\}$.

Proof: Let $A \subseteq \mathbb{R}$, $f: A \to \mathbb{R}$, $r \in \mathbb{R}$, $\lim_{x \to r}[f(x)] = L$, and let $\epsilon = 1$. Then there is $\delta > 0$ such that $0 < |x - r| < \delta$ implies $|f(x) - L| < 1$, or $-1 < f(x) - L < 1$, or $L - 1 < f(x) < L + 1$. Let $a = r - \delta$, $b = r + \delta$, and $M = \max\{|L - 1|, |L + 1|\}$. If $x \in (a, b) \setminus \{r\}$, then $r - \delta < x < r + \delta$ and $x \neq r$. So, $0 < |x - r| < \delta$. Therefore, $L - 1 < f(x) < L + 1$. Since $M \geq |L - 1| \geq 1 - L$, we have $-M \leq L - 1$. Also, $M \geq |L + 1| \geq L + 1$. So, we have $-M < f(x) < M$, or equivalently, $|f(x)| < M$. Since $x \in (a, b) \setminus \{r\}$ was arbitrary, $|f(x)| < M$ for all $x \in (a, b) \setminus \{r\}$. $\square$

14. Let $A \subseteq \mathbb{R}$, let $f, g, h: A \to \mathbb{R}$, let $r \in \mathbb{R}$, let $f(x) \leq g(x) \leq h(x)$ for all $x \in A \setminus \{r\}$, and suppose that $\lim_{x \to r}[f(x)] = \lim_{x \to r}[h(x)] = L$. Prove that $\lim_{x \to r}[g(x)] = L$. This result is known as the **Squeeze Theorem**.

Proof: Let $\epsilon > 0$. Since $\lim_{x \to r}[f(x)] = L$, there is $\delta_1 > 0$ such that $0 < |x - r| < \delta_1$ implies $|f(x) - L| < \epsilon$. Since $\lim_{x \to r}[h(x)] = L$, there is $\delta_2 > 0$ such that $0 < |x - r| < \delta_2$ implies $|h(x) - L| < \epsilon$. Let $\delta = \min\{\delta_1, \delta_2\}$ and let $0 < |x - r| < \delta$. Then $0 < |x - r| < \delta_1$, so that $|f(x) - L| < \epsilon$, or equivalently, $-\epsilon < f(x) - L < \epsilon$, or $L - \epsilon < f(x) < L + \epsilon$. We will need only that $L - \epsilon < f(x)$. Similarly, we have $0 < |x - r| < \delta_2$, so that $|h(x) - L| < \epsilon$, or equivalently, $-\epsilon < h(x) - L < \epsilon$, or $L - \epsilon < h(x) < L + \epsilon$. We will need only that $h(x) < L + \epsilon$. Now, we have $L - \epsilon < f(x) \leq g(x) \leq h(x) < L + \epsilon$. So, $-\epsilon < g(x) - L < \epsilon$, or equivalently, $|g(x) - L| < \epsilon$. Since $\epsilon > 0$ was arbitrary, $\lim_{x \to r}[g(x)] = L$. $\square$

LEVEL 5

15. Let $A \subseteq \mathbb{R}$, let $f, g: A \to \mathbb{R}$ such that $g(x) \neq 0$ for all $x \in A$, let $r \in \mathbb{R}$, and suppose that $\lim_{x \to r}[f(x)]$ and $\lim_{x \to r}[g(x)]$ are both finite real numbers such that $\lim_{x \to r}[g(x)] \neq 0$. Prove that
$$\lim_{x \to r}\left[\frac{f(x)}{g(x)}\right] = \frac{\lim_{x \to r} f(x)}{\lim_{x \to r} g(x)}.$$

Proof: Suppose that $\lim_{x \to r}[f(x)] = L$ and $\lim_{x \to r}[g(x)] = K$, and let $\epsilon > 0$. Since $\lim_{x \to r}[g(x)] = K$, there is $\delta_1 > 0$ such that $0 < |x - r| < \delta_1$ implies $|g(x) - K| < \frac{|K|}{2}$. Now, $|g(x) - K| < \frac{|K|}{2}$ is equivalent to $-\frac{|K|}{2} < g(x) - K < \frac{|K|}{2}$, or by adding K, $K - \frac{|K|}{2} < g(x) < K + \frac{|K|}{2}$. If $K > 0$, we have $\frac{K}{2} < g(x) < \frac{3K}{2}$. If $K < 0$, we have $\frac{3K}{2} < g(x) < \frac{K}{2}$. In both cases, we have $\frac{|K|}{2} < |g(x)| < \frac{3|K|}{2}$. Let $M = \frac{|K|}{2}$. Then $|g(x)| > M$, and so, $\frac{1}{|g(x)|} < \frac{1}{M}$.

Now, since $\lim_{x \to r}[f(x)] = L$, there is $\delta_2 > 0$ such that $0 < |x - r| < \delta_2$ implies $|f(x) - L| < \frac{M|K|\epsilon}{|K| + |L|}$. Since $\lim_{x \to r}[g(x)] = K$, there is $\delta_3 > 0$ such that $0 < |x - r| < \delta_3$ implies $|g(x) - K| < \frac{M|K|\epsilon}{|K| + |L|}$. Let $\delta = \min\{\delta_1, \delta_2, \delta_3\}$ and suppose that $0 < |x - r| < \delta$. Then since $\delta \leq \delta_1$, $\frac{1}{|g(x)|} < \frac{1}{M}$. Since $\delta \leq \delta_2$,

$|f(x) - L| < \frac{M|K|\epsilon}{|K| + |L|}$. Since $\delta \leq \delta_3$, $|g(x) - K| < \frac{M|K|\epsilon}{|K| + |L|}$. By the Triangle Inequality (and SACT), we have

$$\left|\frac{f(x)}{g(x)} - \frac{L}{K}\right| = \left|\frac{Kf(x) - Lg(x)}{Kg(x)}\right| = \left|\frac{Kf(x) - KL + KL - Lg(x)}{Kg(x)}\right| = \left|\frac{Kf(x) - KL}{Kg(x)} + \frac{KL - Lg(x)}{Kg(x)}\right|$$

$$\leq \left|\frac{Kf(x) - KL}{Kg(x)}\right| + \left|\frac{KL - Lg(x)}{Kg(x)}\right| = \left|\frac{f(x) - L}{g(x)}\right| + \left|\frac{L}{K}\right|\left|\frac{K - g(x)}{g(x)}\right| = \left|\frac{f(x) - L}{g(x)}\right| + \left|\frac{L}{K}\right|\left|\frac{g(x) - K}{g(x)}\right|$$

$$= \frac{1}{|g(x)|}\left(|f(x) - L| + \left|\frac{L}{K}\right||g(x) - K|\right) < \frac{1}{M}\left(\frac{M|K|\epsilon}{|K| + |L|} + \left|\frac{L}{K}\right|\frac{M|K|\epsilon}{|K| + |L|}\right) = \frac{1}{M}\cdot\frac{M|K|\epsilon}{|K| + |L|}\left(1 + \left|\frac{L}{K}\right|\right)$$

$$= \frac{|K|\epsilon}{|K| + |L|}\left(\frac{|K| + |L|}{|K|}\right) = \epsilon.$$

So, $\lim\limits_{x \to r}\left[\frac{f(x)}{g(x)}\right] = \frac{L}{K} = \frac{\lim\limits_{x \to r}[f(x)]}{\lim\limits_{x \to r}[g(x)]}.$ □

16. Give a reasonable equivalent definition for each of the following limits (like what was done in Theorem 8.20). r and L are finite real numbers.

  (i) $\lim\limits_{x \to r} f(x) = -\infty$

  (ii) $\lim\limits_{x \to +\infty} f(x) = L$

  (iii) $\lim\limits_{x \to -\infty} f(x) = L$

  (iv) $\lim\limits_{x \to +\infty} f(x) = +\infty$

  (v) $\lim\limits_{x \to +\infty} f(x) = -\infty$

  (vi) $\lim\limits_{x \to -\infty} f(x) = +\infty$

  (vii) $\lim\limits_{x \to -\infty} f(x) = -\infty$

Equivalent definitions:

  (i) $\lim\limits_{x \to r} f(x) = -\infty$ if and only if $\forall M > 0\ \exists \delta > 0\ (0 < |x - r| < \delta \to f(x) < -M).$

  (ii) $\lim\limits_{x \to +\infty} f(x) = L$ if and only if $\forall \epsilon > 0\ \exists K > 0\ (x > K \to |f(x) - L| < \epsilon).$

  (iii) $\lim\limits_{x \to -\infty} f(x) = L$ if and only if $\forall \epsilon > 0\ \exists K > 0\ (x < -K \to |f(x) - L| < \epsilon).$

  (iv) $\lim\limits_{x \to +\infty} f(x) = +\infty$ if and only if $\forall M > 0\ \exists K > 0\ (x > K \to f(x) > M).$

  (v) $\lim\limits_{x \to +\infty} f(x) = -\infty$ if and only if $\forall M > 0\ \exists K > 0\ (x > K \to f(x) < -M).$

  (vi) $\lim\limits_{x \to -\infty} f(x) = +\infty$ if and only if $\forall M > 0\ \exists K > 0\ (x < -K \to f(x) > M).$

  (vii) $\lim\limits_{x \to -\infty} f(x) = -\infty$ if and only if $\forall M > 0\ \exists K > 0\ (x < -K \to f(x) < -M).$

17. Let $f(x) = -x^2 + x + 1$. Use the $M - K$ definition of an infinite limit (that you came up with in Problem 16) to prove $\lim\limits_{x \to +\infty} f(x) = -\infty.$

Proof: Let $M > 0$ and let $K = \frac{1}{2} + \sqrt{M + \frac{5}{4}}$. Suppose that $x > K$. Then $x - \frac{1}{2} > \sqrt{M + \frac{5}{4}}$, and so, $\left(x - \frac{1}{2}\right)^2 > M + \frac{5}{4}$. So, $x^2 - x + \frac{1}{4} > M + \frac{5}{4}$. Thus, $x^2 - x - 1 > M$. So, $-x^2 + x + 1 < -M$. That is, $f(x) < -M$. So, $\lim\limits_{x \to +\infty} g(x) = -\infty$. $\square$

18. Give a reasonable definition for each of the following limits (like what was done in Theorem 8.22). r and L are finite real numbers.

    (i) $\lim\limits_{x \to r^-} f(x) = L$

    (ii) $\lim\limits_{x \to r^+} f(x) = +\infty$

    (iii) $\lim\limits_{x \to r^+} f(x) = -\infty$

    (iv) $\lim\limits_{x \to r^-} f(x) = +\infty$

    (v) $\lim\limits_{x \to r^-} f(x) = -\infty$

Definitions:

    (i) $\lim\limits_{x \to r^-} f(x) = L$ if and only if $\forall \epsilon > 0 \, \exists \delta > 0 \, (-\delta < x - r < 0 \to |f(x) - L| < \epsilon)$.

    (ii) $\lim\limits_{x \to r^+} f(x) = +\infty$ if and only if $\forall M > 0 \, \exists \delta > 0 \, (0 < x - r < \delta \to f(x) > M)$.

    (iii) $\lim\limits_{x \to r^+} f(x) = -\infty$ if and only if $\forall M > 0 \, \exists \delta > 0 \, (0 < x - r < \delta \to f(x) < -M)$.

    (iv) $\lim\limits_{x \to r^-} f(x) = +\infty$ if and only if $\forall M > 0 \, \exists \delta > 0 \, (-\delta < x - r < 0 \to f(x) > M)$.

    (v) $\lim\limits_{x \to r^-} f(x) = -\infty$ if and only if $\forall M > 0 \, \exists \delta > 0 \, (-\delta < x - r < 0 \to f(x) < -M)$.

19. Use the $M - \delta$ definition of a one-sided limit (that you came up with in Problem 18) to prove that $\lim\limits_{x \to 3^-} \frac{1}{x-3} = -\infty$.

Proof: Let $M > 0$ and let $\delta = \frac{1}{M}$. If $-\delta < x - 3 < 0$, then $-\frac{1}{M} < x - 3 < 0$, and so, we have $\frac{1}{x-3} < -M$. Since $M > 0$ was arbitrary, $\lim\limits_{x \to 3^-} \frac{1}{x-3} = -\infty$. $\square$

20. Let $f(x) = \frac{x+1}{(x-1)^2}$. Prove that

    (i) $\lim\limits_{x \to +\infty} f(x) = 0$.

    (ii) $\lim\limits_{x \to 1^+} f(x) = +\infty$.

Proofs:

(i) Let $\epsilon > 0$ and let $K = \max\left\{2, 1 + \frac{3}{\epsilon}\right\}$. Let $x > K$. Then $x - 1 > 1 + \frac{3}{\epsilon} - 1 = \frac{3}{\epsilon}$, and therefore, $\frac{1}{x-1} < \frac{\epsilon}{3}$. Also, since $x > 2$, $(x-1)^2 - (x-1) = (x-1)(x-1-1) = (x-1)(x-2) > 0$ (because $x - 1 > 2 - 1 = 1 > 0$ and $x - 2 > 2 - 2 > 0$). Thus, $(x-1)^2 > x - 1$, and so, $\frac{1}{(x-1)^2} < \frac{1}{x-1} < \frac{\epsilon}{3}$. It follows from the triangle inequality (and SACT) that

$$\left|\frac{x+1}{(x-1)^2} - 0\right| = \left|\frac{x-1+2}{(x-1)^2}\right| = \left|\frac{x-1}{(x-1)^2} + \frac{2}{(x-1)^2}\right| = \left|\frac{1}{x-1} + \frac{2}{(x-1)^2}\right|$$

$$\leq \left|\frac{1}{x-1}\right| + \left|\frac{2}{(x-1)^2}\right| = \frac{1}{x-1} + 2\frac{1}{(x-1)^2} < \frac{\epsilon}{3} + 2 \cdot \frac{\epsilon}{3} = 3 \cdot \frac{\epsilon}{3} = \epsilon.$$

So, $\lim\limits_{x \to +\infty} f(x) = 0$. □

(ii) Let $M > 0$ and let $\delta = \min\left\{1, \frac{3}{M}\right\}$. If $0 < x - 1 < \delta$, then $0 < x - 1 < \frac{3}{M}$, and so, we have $\frac{1}{x-1} > \frac{M}{3}$. Since $0 < x - 1 < 1$, $(x-1)^2 < x - 1$, and so, $\frac{1}{(x-1)^2} > \frac{1}{x-1}$. So, we have

$$\frac{x+1}{(x-1)^2} = \frac{x-1+2}{(x-1)^2} = \frac{x-1}{(x-1)^2} + \frac{2}{(x-1)^2} = \frac{1}{x-1} + \frac{2}{(x-1)^2}$$

$$> \frac{1}{x-1} + \frac{2}{x-1} = \frac{3}{x-1} > 3 \cdot \frac{M}{3} = M.$$

So, $\lim\limits_{x \to 1^+} f(x) = +\infty$. □

21. Let $f: \mathbb{R} \to \mathbb{R}$ be defined by $f(x) = \begin{cases} 0 & \text{if } x \text{ is rational.} \\ 1 & \text{if } x \text{ is irrational.} \end{cases}$ Prove that for all $r \in \mathbb{R}$, $\lim\limits_{x \to r}[f(x)]$ does not exist.

Proof: Let $r \in \mathbb{R}$, let $\epsilon = \frac{1}{2}$, and let $\delta > 0$. By the Density Theorem (Theorem 6.17) and Problem 19 from Problem Set 6, there is a rational number x and an irrational number y such that $r < x, y < r + \delta$. So, we have $0 < |x - r| < \delta$ and $0 < |y - r| < \delta$. We also have $f(x) = 0$ and $f(y) = 1$. Let $L \in \mathbb{R}$. If $\lim\limits_{x \to r}[f(x)] = L$, then $|f(x) - L| < \frac{1}{2}$ and $|f(y) - L| < \frac{1}{2}$. But then by the Triangle Inequality, we would have

$$|f(x) - f(y)| = |f(x) - L + L - f(y)| \leq |f(x) - L| + |L - f(y)| < \frac{1}{2} + \frac{1}{2} = 1.$$

However, $|f(x) - f(y)| = |1 - 0| = 1$. Since $1 < 1$ is false, $\lim\limits_{x \to r}[f(x)]$ does not equal L. Since $L \in \mathbb{R}$ was arbitrary, $\lim\limits_{x \to r}[f(x)]$ does not exist. □

22. Let $f, g: \mathbb{R} \to \mathbb{R}$ be defined by $f(x) = \cos x$ and $g(x) = \sin x$. Prove that f and g are uniformly continuous on $\mathbb{R}$. Hint: Use the fact that the least distance between two points is a straight line.

Proof: Let $\epsilon > 0$ and let $\delta = \min\{\epsilon, 2\pi\}$. Let $x, y \in \mathbb{R}$ with $|x - y| < \delta$. Suppose that $W(x) = (a, b)$ and $W(y) = (c, d)$. The arc length along the unit circle between (a, b) and (c, d) is $|x - y|$ and the straight-line distance between (a, b) and (c, d) is $\sqrt{(a - c)^2 + (b - d)^2}$. Thus,

$$|\cos x - \cos y| = |a - c| \leq \sqrt{(a - c)^2 + (b - d)^2}$$
$$\leq |x - y|.$$

$$|\sin x - \sin y| = |b - d| \leq \sqrt{(a - c)^2 + (b - d)^2}$$
$$\leq |x - y|.$$

Therefore, we have $|\cos x - \cos y| \leq |x - y| < \delta \leq \epsilon$ and $|\sin x - \sin y| \leq |x - y| < \delta \leq \epsilon$. It follows that f and g are uniformly continuous on $\mathbb{R}$. $\square$

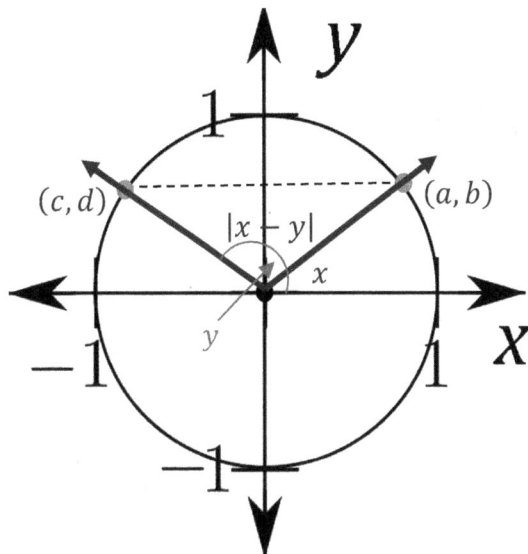

23. Prove each of the following:

(i) $\displaystyle\lim_{x \to 0} \frac{\sin x}{x} = 1$

(ii) $\displaystyle\lim_{x \to 0} \frac{\cos x - 1}{x} = 0$

Proofs:

(i) By Problem 17 from Problem Set 3, for $x \in \left(-\frac{\pi}{2}, 0\right) \cup \left(0, \frac{\pi}{2}\right)$, $\cos x \leq \frac{\sin x}{x} \leq 1$. By Problem 22, $\cos x$ is continuous at $x = 0$, and so, by Problem 5, $\lim_{x \to 0} \cos x = \cos 0 = 1$. By Problem 2, $\lim_{x \to 0} 1 = 1$. By the Squeeze Theorem (Problem 14), $\lim_{x \to 0} \frac{\sin x}{x} = 1$. $\square$

(ii) $\displaystyle\lim_{x \to 0} \frac{\cos x - 1}{x} = \lim_{x \to 0} \frac{(\cos x - 1)(\cos x + 1)}{x(\cos x + 1)} = \lim_{x \to 0} \frac{\cos^2 x - 1}{x(\cos x + 1)} = \lim_{x \to 0} \frac{-\sin^2 x}{x(\cos x + 1)} = -\lim_{x \to 0} \left(\frac{\sin x}{x} \cdot \frac{\sin x}{\cos x + 1}\right)$. By Theorem 8.14, this is equal to $-\left(\lim_{x \to 0} \frac{\sin x}{x} \cdot \lim_{x \to 0} \frac{\sin x}{\cos x + 1}\right)$. By part (i) above, this is equal to $-\left(1 \cdot \lim_{x \to 0} \frac{0}{2}\right) = -(1 \cdot 0) = 0$. $\square$

Problem Set 9

LEVEL 1

1. Let $A \subseteq \mathbb{R}$ and let $f: A \to \mathbb{R}$. Provide a counterexample showing that each of the following is false.

   (i)   If A is closed and f is continuous, then f is bounded.

   (ii)   If A is bounded and f is continuous, then f is bounded.

   (iii) If A is closed and bounded, then f is bounded.

   (iv) If A is a closed and f is continuous then $f[A]$ is closed.

   (v)   If A is open and f is continuous, then $f[A]$ is open.

Solutions:

(i)   Let $A = [0, \infty)$ and define $f: A \to \mathbb{R}$ by $f(x) = x$. Then A is closed and f is continuous, but f is not bounded.

(ii)   Let $A = (0, 1)$ and define $f: A \to \mathbb{R}$ by $f(x) = \frac{1}{x}$. Then A is bounded and f is continuous, but f is not bounded.

(iii) Let $A = [0, 1]$ and define $f: A \to \mathbb{R}$ by $f(x) = \begin{cases} 1 & \text{if } x = 0. \\ \frac{1}{x} & \text{if } x \neq 0. \end{cases}$ Then A is closed and bounded, but f is not bounded.

(iv) Let $A = [0, \infty)$ and define $f: A \to \mathbb{R}$ by $f(x) = \frac{1}{x^2+1}$. Then A is closed and f is continuous on A, but $f[A] = (0, 1]$, which is not closed.

(v)   Let $A = (-1, 1)$ and define $f: A \to \mathbb{R}$ by $f(x) = \frac{1}{x^2+1}$. Then A is open and f is continuous on A, but $f[A] = \left(\frac{1}{2}, 1\right]$, which is not open.

2. Let I be a closed bounded interval and let J be an open or half-open interval. Prove that I and J are not homeomorphic.

Proof: Suppose toward contradiction that $f: I \to J$ is a homeomorphism. By the Heine-Borel Theorem, I is compact. By Theorem 9.4, $f[I] = J$ is compact. But then, by the Heine-Borel Theorem again, J is closed, which is a contradiction.   □

LEVEL 2

3. Let $A \subseteq \overline{A} \subseteq B \subseteq \mathbb{R}$ and let $f, g: B \to \mathbb{R}$ be continuous. Suppose that $f(x) = g(x)$ for all $x \in A$. Prove that $f(x) = g(x)$ for all $x \in \overline{A}$.

90

Proof: Let $x \in \overline{A}$. Then there is a sequence (x_n) in A such that $x_n \to x$ (simply choose $x_n \in N_{\frac{1}{n}}(x) \cap A$ for each $n \in \mathbb{N}$). Since f is continuous, $f(x_n) \to f(x)$. Since g is continuous, $f(x_n) = g(x_n) \to g(x)$. So, $f(x_n)$ converges to both $f(x)$ and $g(x)$. By the Note following Example 8.23, $f(x) = g(x)$. $\qquad\square$

LEVEL 3

4. Let $A \subseteq \mathbb{R}$ and let $f: A \to \mathbb{R}$. Prove that f is continuous on A if and only if for each closed set $C \subseteq \mathbb{R}$, there is a closed set $B \subseteq \mathbb{R}$ such that $f^{-1}[C] = B \cap A$.

Proof: Let $f: A \to \mathbb{R}$. First assume that f is continuous on A and let $C \subseteq \mathbb{R}$ be closed. Then $\mathbb{R} \setminus C$ is open. Since f is continuous on A, by Theorem 9.3, there is an open set $U \subseteq \mathbb{R}$ such that $f^{-1}[\mathbb{R} \setminus C] = U \cap A$. We first show that $f^{-1}[\mathbb{R} \setminus C] = A \setminus f^{-1}[C]$. To see this, note that $x \in f^{-1}[\mathbb{R} \setminus C]$ if and only if $f(x) \in \mathbb{R} \setminus C$ if and only if $f(x) \in \mathbb{R}$ and $f(x) \notin C$ if and only if $x \in A$ and $x \notin f^{-1}[C]$ if and only if $x \in A \setminus f^{-1}[C]$. It follows that $A \setminus f^{-1}[C] = U \cap A$. Let $B = \mathbb{R} \setminus U$. Then B is closed. We complete the argument by showing that $f^{-1}[C] = B \cap A$. To see this, note that $x \in f^{-1}[C]$ if and only if $x \notin A \setminus f^{-1}[C] = U \cap A$. Since x must be in A, $x \in f^{-1}[C]$ if and only if $x \notin U$ and $x \in A$ if and only if $x \in B$ and $x \in A$ if and only if $x \in B \cap A$.

Conversely, assume that for each closed $C \subseteq \mathbb{R}$, there is a closed set $B \subseteq \mathbb{R}$ such that $f^{-1}[C] = B \cap A$. Let $U \subseteq \mathbb{R}$ be open. Then $\mathbb{R} \setminus U$ is closed. By our assumption, there is a closed set $B \subseteq \mathbb{R}$ such that $f^{-1}[\mathbb{R} \setminus U] = B \cap A$. By the same argument given in the first paragraph above, we have $f^{-1}[\mathbb{R} \setminus U] = A \setminus f^{-1}[U]$. It follows that $A \setminus f^{-1}[U] = B \cap A$. Let $V = \mathbb{R} \setminus B$. Then V is open. We complete the argument by showing that $f^{-1}[U] = V \cap A$. To see this, note that $x \in f^{-1}[U]$ if and only if $x \notin A \setminus f^{-1}[U] = B \cap A$. Since x must be in A, $x \in f^{-1}[U]$ if and only if $x \notin B$ and $x \in A$ if and only if $x \in V$ and $x \in A$ if and only if $x \in V \cap A$. $\qquad\square$

5. Prove that the image of a Cauchy sequence under a uniformly continuous function is a Cauchy sequence. If we replace "uniformly continuous" by "continuous," is the result still true?

Proof: Let $A \subseteq \mathbb{R}$, let $f: A \to B$ be uniformly continuous, let (x_n) be a Cauchy sequence in A, and let $\epsilon > 0$. Since f is uniformly continuous, there is $\delta > 0$ such that for all $x, y \in A$, $|x - y| < \delta \to |f(x) - f(y)| < \epsilon$. Since (x_n) is a Cauchy sequence, there is $K \in \mathbb{N}$ such that $m \geq n > K$ implies $|x_m - x_n| < \delta$. Now, assume that $m \geq n > K$. Then we have $|x_m - x_n| < \delta$. Therefore, $|f(x_m) - f(x_n)| < \epsilon$. It follows that $(f(x_n))$ is a Cauchy sequence in $\mathbb{R}$. $\qquad\square$

If we replace "uniformly continuous" by "continuous," the result is false. Define $f: (0, 1) \to \mathbb{R}$ by $f(x) = \frac{1}{x}$. Then $\left(\frac{1}{n}\right)$ is a Cauchy sequence in $(0, 1)$, but $f\left(\frac{1}{n}\right) = n$, and (n) is not a Cauchy sequence.

6. Let $S \subseteq \mathbb{R}$, let A and B be disjoint closed subsets of S, and let a and b be real numbers with $a < b$. Use Urysohn's Lemma (see Problem 13 below) to prove that there is a continuous function $f: S \to [a, b]$ such that $f[A] = \{a\}$ and $f[B] = \{b\}$.

Proof: Let $S \subseteq \mathbb{R}$, let A and B be disjoint closed subsets of S, and let a and b be real numbers with $a < b$. By Urysohn's Lemma (Problem 13), there is a continuous function $g: S \to [0,1]$ such that $g[A] = \{0\}$ and $g[B] = \{1\}$. Define $f: S \to [a,b]$ by $f(x) = (b-a)g(x) + a$. Then we have $f[A] = (b-a)g[A] + a = (b-a)(0) + a = a$, $f[B] = (b-a)g[B] + a = (b-a)(1) + a = b$, and f is continuous by a standard continuity argument. $\square$

---

7. Prove that any two open intervals are homeomorphic. (You may exclude $(-\infty, \infty)$.)

Proof: In the solution to part (iii) of Problem 7 from Problem Set 5, we saw that if $a, b \in \mathbb{R}$ with $a \neq b$, then the function $h: (0,1) \to (a,b)$ defined by $h(x) = (b-a)x + a$ is a bijection. By Problem 6 in Problem Set 8, h is continuous. Also, $h^{-1}(x) = \frac{1}{b-a}(x-a)$ is continuous for the same reason. It follows that h is a homeomorphism. Since compositions of bijective functions are bijective (Corollary 3.7) and compositions of continuous functions are continuous (Problem 8 from Problem Set 7), it follows that any two bounded open intervals are homeomorphic.

Again, in part (iii) of Problem 7 from Problem Set 5, we saw that the function $g: (0, \infty) \to (0,1)$ defined by $g(x) = \frac{1}{x^2 + 1}$ is a bijection. If $a \in (0, \infty)$, then by Problems 6 and 15 in Problem Set 8, $\lim_{x \to a} g(x) = \frac{1}{a^2 + 1} = g(a)$. By Problem 5 in Problem Set 8, g is continuous. Also, $g^{-1}(x) = \sqrt{\frac{1}{x} - 1}$. By part (ii) of Problem 10 in Problem Set 8, as well as other results already mentioned, g^{-1} is continuous. It follows that g is a homeomorphism.

The function $k: (0, \infty) \to (-\infty, 0)$ defined by $f(x) = -x$ is easily seen to be a homeomorphism. Using the fact that compositions of bijective functions are bijective (Corollary 3.7) and compositions of continuous functions are continuous (Problem 8 from Problem Set 7), we now have that any two open intervals I and J are homeomorphic, as long as $I \neq (-\infty, \infty)$ and $J \neq (-\infty, \infty)$. $\square$

Note: It is true that $\mathbb{R} = (-\infty, \infty)$ is homeomorphic to every other open interval as well. For example, the function $f: \mathbb{R} \to (0, \infty)$ defined by $f(x) = 2^x$ is a homeomorphism. However, to prove this requires a formal definition of 2^x. This definition and the theory necessary to prove this result will be provided in Lesson 12.

LEVEL 4

8. Let $A \subseteq \mathbb{R}$ and let $f: A \to \mathbb{R}$. Prove that f is continuous on A if and only if for each $X \subseteq A$, $f[\overline{X} \cap A] \subseteq \overline{f[X]}$.

Proof: Let $A \subseteq \mathbb{R}$ and let $f: A \to \mathbb{R}$. First assume that f is continuous on A. Let $X \subseteq A$ and let $y \in f[\overline{X} \cap A]$. Then there is $x \in \overline{X} \cap A$ with $y = f(x)$. Now, let $V \subseteq \mathbb{R}$ be an open interval containing $f(x)$. By Theorem 9.1 ($1 \to 4$), there is an open set $U \subseteq \mathbb{R}$ with $x \in U$ such that $f[U \cap A] \subseteq V$. By part 5 of Theorem 7.25, there is $z \in U \cap X$ (Check this carefully!), Since $X \subseteq A$, $f[U \cap X] \subseteq f[U \cap A]$. Therefore, $f(z) \in f[U \cap X] \subseteq f[U \cap A] \subseteq V$. Also, since $z \in X$, we have $f(z) \in f[X]$. Once again, by part 5 of Theorem 7.25, $y = f(x) \in \overline{f[X]}$. Since $y \in f[\overline{X} \cap A]$ was arbitrary, $f[\overline{X} \cap A] \subseteq \overline{f[X]}$.

Conversely, assume that for each $X \subseteq A$, $f[\overline{X} \cap A] \subseteq \overline{f[X]}$. Let C be a closed set in $\mathbb{R}$ and let $D = f^{-1}[C]$. By Problem 5 from Problem Set 3, $f[D] = f[f^{-1}[C]] \subseteq C$. Now, if $x \in \overline{D} \cap A$, then we have $f(x) \in f[\overline{D} \cap A]$. By our assumption, $f[\overline{D} \cap A] \subseteq \overline{f[D]}$. So, $f(x) \in \overline{f[D]} \subseteq \overline{C} = C$ (because C is closed in $\mathbb{R}$). So, $f(x) \in C$, and therefore, $x \in f^{-1}[C] = D$. Since $x \in \overline{D} \cap A$ was arbitrary, $\overline{D} \cap A \subseteq D$. Since $D \subseteq \overline{D}$ is always true (by part 1 of Theorem 7.25) and $D \subseteq A$ (because $x \in D = f^{-1}[C]$ implies that $f(x) \in C$, which implies that $x \in \text{dom } f = A$), we have $D \subseteq \overline{D} \cap A$. Since both $\overline{D} \cap A \subseteq D$ and $D \subseteq \overline{D} \cap A$, it follows that $f^{-1}[C] = D = \overline{D} \cap A$. Since $\overline{D}$ is closed in $\mathbb{R}$, by problem 4 above, f is continuous on A. $\qquad\square$

9. Let $A \subseteq \mathbb{R}$ be compact and let $f: A \to \mathbb{R}$ be continuous and injective. Prove that f is a homeomorphism.

Proof: We need to show that $f^{-1}: f[A] \to A$ is continuous. Let $V \subseteq \mathbb{R}$ be open. Then $\mathbb{R} \setminus V$ is closed. Since A is compact, by the Heine-Borel Theorem, A is closed. Therefore, $A \cap (\mathbb{R} \setminus V)$ is closed. By Problem 14 from Problem Set 7, $A \cap (\mathbb{R} \setminus V) = A \setminus V$ is compact. By Theorem 9.4, $f[A \setminus V]$ is compact. Once again, the Heine-Borel Theorem tells us that $f[A \setminus V]$ is closed. Since f is injective, $f[A \setminus V] = f[A] \setminus f[V]$ (Check this carefully). So, $\mathbb{R} \setminus (f[A] \setminus f[V])$ is open. We now show that $f[V] = (\mathbb{R} \setminus (f[A] \setminus f[V])) \cap f[A]$. To see this, note that $x \in (\mathbb{R} \setminus (f[A] \setminus f[V])) \cap f[A]$ if and only if $x \notin f[A] \setminus f[V]$ and $x \in f[A]$ if and only if we have both $x \in f[V]$ or $x \notin f[A]$ as well as $x \in f[A]$ if and only if $x \in f[V] \cap f[A] = f[V]$. By Theorem 9.3, f^{-1} is continuous. $\qquad\square$

LEVEL 5

10. Let $A \subseteq \mathbb{R}$ be compact and let $f: A \to \mathbb{R}$ be continuous on A. Prove that f is uniformly continuous on A.

Proof: Let $A \subseteq \mathbb{R}$ be compact, let $f: A \to \mathbb{R}$ be continuous on A, and let $\epsilon > 0$. By the continuity of f, for each $a \in A$, there is a positive number δ_a such that $\forall b \in A \left(|a - b| < \delta_a \to |f(a) - f(b)| < \frac{\epsilon}{2} \right)$. For each $a \in A$, let $B_a = \left\{ b \in A \mid |a - b| < \frac{\delta_a}{2} \right\}$. Then $\mathcal{C} = \{ B_a \mid a \in A \}$ is an open cover of A. Since A is compact, there is a finite subcover $\mathcal{D} \subseteq \mathcal{C}$. Let $\delta = \min \left\{ \frac{\delta_a}{2} \mid B_a \in \mathcal{D} \right\}$. Since $\mathcal{D}$ is finite, $\delta > 0$. Let $a, b \in A$ with $|a - b| < \delta$. Let $B_c \in \mathcal{D}$ with $a \in B_c$. Then $|a - c| < \frac{\delta_c}{2}$. Also,

$$|b - c| \leq |b - a + a - c| \leq |b - a| + |a - c| < \delta + \frac{\delta_c}{2} \leq \delta_c.$$

So, $|f(a) - f(b)| = |f(a) - f(c) + f(c) - f(b)| \leq |f(a) - f(c)| + |f(c) - f(b)| < \frac{\epsilon}{4} + \frac{\epsilon}{2} < \epsilon.$ $\quad\square$

11. Let $C \subseteq A \subseteq \mathbb{R}$ with C dense in A and suppose that $f: C \to \mathbb{R}$ is uniformly continuous. Prove that f can be extended uniquely to a uniformly continuous function $g: A \to \mathbb{R}$.

Proof: Let $C \subseteq A \subseteq \mathbb{R}$ with C dense in A and let $f: C \to \mathbb{R}$ be uniformly continuous. If $C = A$, we can let $g = f$. So, assume that $C \neq A$. If $x \in C$, let $g(x) = f(x)$. Now, let $x \in A \setminus C$. Since C is dense in A, $\overline{C} = A$. Therefore, there is a sequence (x_n) in C such that $x_n \to x$. By Theorem 8.25, (x_n) is a Cauchy sequence in C. By Problem 5 above, $\big(f(x_n)\big)$ is a Cauchy sequence in $\mathbb{R}$. By Theorem 8.28, $\big(f(x_n)\big)$ converges to some point y in $\mathbb{R}$. Let $g(x) = y$. If (z_n) is another Cauchy sequence in C such that $z_n \to x$, then $|x_n - z_n| \to 0$. Since f is uniformly continuous, $|f(x_n) - f(z_n)| \to 0$. Therefore, $f(z_n) \to y = g(x)$.

We need to show that g is uniformly continuous. To this end, let $\epsilon > 0$. Since f is uniformly continuous, there is $\delta > 0$ such that for all $x, z \in C$, $|x - z| < \delta \to |f(x) - f(z)| < \frac{\epsilon}{2}$. Suppose that $a, b \in A$ with $|a - b| < \delta$. Let $(a_n), (b_n)$ be sequences in C such that $a_n \to a$ and $b_n \to b$. We then have that $|a_n - b_n| = |a_n - a + a - b + b - b_n| \leq |a_n - a| + |a - b| + |b - b_n|$. Since $|a_n - a| \to 0$, $|b - b_n| \to 0$, and $|a - b| < \delta$, we can find $K \in \mathbb{N}$ such that for all $n > K$, $|a_n - b_n| < \delta$. So, $n > K \to |f(a_n) - f(b_n)| < \frac{\epsilon}{2}$. From this it is easy to show that $|g(a) - g(b)| \leq \frac{\epsilon}{2} < \epsilon$.

The proof that g is unique is straightforward. $\qquad\qquad\qquad\qquad\qquad\qquad\qquad\qquad$ $\square$

Problem Set 10

LEVEL 1

1. Let $r \in \mathbb{R}$ and define $f: \mathbb{R} \to \mathbb{R}$ by $f(x) = r$. Prove that for any $a \in \mathbb{R}$, $f'(a) = 0$.

Proof:

$$f'(a) = \lim_{x \to a} \frac{f(x) - f(a)}{x - a} = \lim_{x \to a} \frac{r - r}{x - a} = \lim_{x \to a} \frac{0}{x - a} = 0.$$

$\square$

2. Define a continuous function $f: \mathbb{R} \to \mathbb{R}$ that is not differentiable on $\mathbb{R}$.

Solution: Let $f(x) = |x|$. Then f is continuous on $\mathbb{R}$, but f is not differentiable at 0. To see that f is not differentiable at 0, observe that

$$f'(0) = \lim_{x \to 0} \frac{f(x) - f(0)}{x - 0} = \lim_{x \to 0} \frac{|x| - 0}{x} = \lim_{x \to 0} \frac{|x|}{x}.$$

Since $\lim_{x \to 0^+} \frac{|x|}{x} = \lim_{x \to 0^+} \frac{x}{x} = \lim_{x \to 0^+} 1 = 1$ and $\lim_{x \to 0^-} \frac{|x|}{x} = \lim_{x \to 0^-} \frac{-x}{x} = \lim_{x \to 0^-} (-1) = -1$, it follows that $f'(0)$ does not exist.

$\square$

3. Compute the derivatives of $\cos x$, $\tan x$, $\sec x$, and $\cot x$.

Solutions:

$$\frac{d}{dx}[\cos x] = \lim_{h \to 0} \frac{\cos(x + h) - \cos x}{h} = \lim_{h \to 0} \frac{\cos x \cos h - \sin x \sin h - \cos x}{h}$$

$$= \lim_{h \to 0} \frac{(\cos x \cos h - \cos x) - \sin x \sin h}{h} = \lim_{h \to 0} \frac{\cos x (\cos h - 1) - \sin x \sin h}{h}$$

$$= \lim_{h \to 0} \frac{\cos x (\cos h - 1)}{h} - \lim_{h \to 0} \frac{\sin x \sin h}{h} = (\cos x) \lim_{h \to 0} \frac{\cos h - 1}{h} - (\sin x) \lim_{h \to 0} \frac{\sin h}{h}$$

$$= (\cos x)(0) - (\sin x)(1) = -\sin x \text{ (by Problem 23 in Problem Set 8).}$$

$$\frac{d}{dx}[\tan x] = \frac{d}{dx}\left[\frac{\sin x}{\cos x}\right] = \frac{(\cos x)(\cos x) - (\sin x)(-\sin x)}{\cos^2 x} = \frac{\cos^2 x + \sin^2 x}{\cos^2 x} = \frac{1}{\cos^2 x} = \sec^2 x.$$

$$\frac{d}{dx}[\sec x] = \frac{d}{dx}\left[\frac{1}{\cos x}\right] = \frac{(\cos x)(0) - (1)(-\sin x)}{\cos^2 x} = \frac{\sin x}{\cos^2 x} = \frac{1}{\cos x} \cdot \frac{\sin x}{\cos x} = \sec x \tan x.$$

$$\frac{d}{dx}[\cot x] = \frac{d}{dx}[(\tan x)^{-1}] = -(\tan x)^{-2} \sec^2 x = -\frac{\cos^2 x}{\sin^2 x} \cdot \frac{1}{\cos^2 x} = -\frac{1}{\sin^2 x} = -\csc^2 x.$$

95

4. Let $A \subseteq \mathbb{R}$, let $k \in \mathbb{R}$, and let $f: A \to \mathbb{R}$ be differentiable at $a \in A$. Prove that the function $kf: A \to \mathbb{R}$ defined by $(kf)(x) = k \cdot f(x)$ is differentiable at a and $(kf)'(a) = k \cdot f'(a)$.

Proof:

$$(kf)'(a) = \lim_{x \to a} \frac{(kf)(x) - (kf)(a)}{x - a} = \lim_{x \to a} \frac{k \cdot f(x) - k \cdot f(a)}{x - a} = \lim_{x \to a} \frac{k(f(x) - f(a))}{x - a}$$

$$= k \cdot \lim_{x \to a} \frac{f(x) - f(a)}{x - a} = k \cdot f'(a).$$

$\square$

5. Suppose that $(a, b) \subseteq A \subseteq \mathbb{R}$, $f, g: A \to \mathbb{R}$ are differentiable on (a, b), and $f'(x) = g'(x)$ for all $x \in (a, b)$. Prove that there is a constant k such that for all $x \in (a, b)$, $g(x) = f(x) + k$.

Proof: Define $h: A \to \mathbb{R}$ by $h(x) = g(x) - f(x)$. Then, for all $x \in (a, b)$, $h'(x) = g'(x) - f'(x) = 0$. Let $c, d \in (a, b)$ with $c < d$. Then f is continuous on $[c, d]$ and differentiable on (c, d). So, by the Mean Value Theorem, there is $e \in (c, d)$ with $\frac{h(d) - h(c)}{d - c} = h'(e) = 0$. It follows that $h(d) = h(c)$. Since $c, d \in (a, b)$ were arbitrary, h is constant on (a, b). So, there is a constant k such that for all $x \in (a, b)$, $h(x) = k$. Therefore, $g(x) - f(x) = k$, or equivalently, $g(x) = f(x) + k$.

$\square$

6. Suppose that $(a, b) \subseteq A \subseteq \mathbb{R}$, $f, g: A \to \mathbb{R}$ are continuous on A and differentiable on (a, b), $\lim_{x \to b^-} f(x) = \lim_{x \to b^-} g(x) = 0$, $g'(x) \neq 0$ for all $x \in (a, b)$, and $\lim_{x \to b^-} \frac{f'(x)}{g'(x)}$ exists. Prove that

$$\lim_{x \to b^-} \frac{f(x)}{g(x)} = \lim_{x \to b^-} \frac{f'(x)}{g'(x)}.$$

This is standard version B of L'Hôpital's Rule (Theorem 10.30).

Proof: If $b \in A$, then since f and g are continuous on A and $\lim_{x \to b^-} f(x) = \lim_{x \to b^-} g(x) = 0$, we must have $f(b) = 0$ and $g(b) = 0$. In general, let $B = A \cup \{b\}$ and define $F, G: B \to \mathbb{R}$ as follows:

$$F(x) = \begin{cases} f(x) & \text{if } x \in A \\ 0 & \text{if } x = b \end{cases} \qquad G(x) = \begin{cases} g(x) & \text{if } x \in A \\ 0 & \text{if } x = b \end{cases}$$

Note that if $b \in A$, then $B = A$, $F = f$, and $G = g$. So, there is no harm in simply assuming that $b \in A$ and $f(b) = g(b) = 0$. It follows that f and g are continuous on $(a, b]$.

For any $x \in (a, b)$, we see that f and g are continuous on $[x, b]$ and differentiable on (x, b). By the Mean Value Theorem (Theorem 10.25), there is $c \in (x, b)$ such that $g'(c) = \frac{g(b) - g(x)}{b - x} = \frac{-g(x)}{b - x}$. Since $g'(c) \neq 0$, we must also have $g(x) \neq 0$. By the Generalized Mean value Theorem (Theorem 10.27), there is $c \in (x, b)$ such that $f(x)g'(c) = g(x)f'(c)$, or equivalently, $\frac{f(x)}{g(x)} = \frac{f'(c)}{g'(c)}$. Since $x < c < b$, as x approaches b from the left, so does c. So, we have

$$\lim_{x \to b^-} \frac{f(x)}{g(x)} = \lim_{x \to b^-} \frac{f'(c)}{g'(c)} = \lim_{c \to b^-} \frac{f'(c)}{g'(c)} = \lim_{x \to b^-} \frac{f'(x)}{g'(x)}.$$

Note that the last equality above is true because we simply changed the name of the variable. □

LEVEL 3

7. Let $A, B \subseteq \mathbb{R}$, suppose that $f: A \to \mathbb{R}$, $g: B \to \mathbb{R}$ are differentiable at $a \in A \cap B$, and assume that $g(a) \neq 0$. Let $C = (A \cap B) \setminus \{x \in B \mid g(x) = 0\}$. Prove that the function $f/g: C \to \mathbb{R}$ defined by $(f/g)(x) = \frac{f(x)}{g(x)}$ is differentiable at a and $(f/g)'(a) = \frac{g(a)f'(a) - f(a)g'(a)}{(g(a))^2}$. This is the quotient rule (Theorem 10.18).

Proof: Using the product rule and chain rule, we have

$$\left(\frac{f}{g}\right)'(x) = \frac{d}{dx}\left[\frac{f(x)}{g(x)}\right] = \frac{d}{dx}\left[f(x)(g(x))^{-1}\right] = f(x)(-1)(g(x))^{-2}g'(x) + (g(x))^{-1}f'(x)$$

$$= -\frac{f(x)g'(x)}{(g(x))^2} + \frac{f'(x)}{g(x)} = -\frac{f(x)g'(x)}{(g(x))^2} + \frac{f'(x)g(x)}{(g(x))^2} = \frac{g(x)f'(x) - f(x)g'(x)}{(g(x))^2}.$$

Substituting a for x gives the desired result. □

8. Suppose that $f(c)$ is a relative minimum for a function $f: A \to \mathbb{R}$. Prove that either $f'(c) = 0$ or f is not differentiable at c.

Proof: Suppose that $f(c)$ is a relative minimum for $f: A \to \mathbb{R}$ and f is differentiable at c. First, let's suppose toward contradiction that $f'(c) > 0$. In other words, we have

$$\lim_{x \to c} \frac{f(x) - f(c)}{x - c} > 0.$$

By Problem 12 from Problem Set 8, there is a deleted neighborhood N of c such that $\frac{f(x) - f(c)}{x - c} > 0$ for all $x \in N$. Now, if $x \in N$ and $x < c$, then $x - c < 0$, and therefore,

$$f(x) - f(c) = \frac{f(x) - f(c)}{x - c}(x - c) < 0.$$

So, for $x \in N$ with $x < c$, we have $f(x) < f(c)$, contradicting our assumption that $f(c)$ is a relative minimum for f. So, it is impossible to have $f'(c) > 0$.

Next, let's suppose toward contradiction that $f'(c) < 0$. Then we have

$$\lim_{x \to c} \frac{f(x) - f(c)}{x - c} < 0.$$

Therefore, it follows that

$$\lim_{x \to c} \frac{f(c) - f(x)}{x - c} = -\left[\lim_{x \to c} \frac{f(x) - f(c)}{x - c}\right] > 0.$$

Again, by Problem 12 from Problem Set 8, there is a deleted neighborhood N of c such that $\frac{f(c)-f(x)}{x-c} > 0$ for all $x \in N$. Now, if $x \in N$ and $x > c$, then $x - c > 0$, and therefore,

$$f(c) - f(x) = \frac{f(c) - f(x)}{x - c}(x - c) > 0.$$

So, for $x \in N$ with $x > c$, we have $f(c) > f(x)$, once again contradicting our assumption that $f(c)$ is a relative minimum for f. So, it is also impossible to have $f'(c) < 0$.

Since f is differentiable at c, we must have $f'(c) = 0$. $\qquad\square$

9. Use the Mean Value Theorem to prove that $-x \leq \sin x \leq x$ for all $x \in [0, \infty)$.

Proof: If $x \geq 1$, then $-x \leq -1$. Since $-1 \leq \sin x \leq 1$, we have $-x \leq -1 \leq \sin x \leq 1 \leq x$.

If $x = 0$, we have $\sin 0 = 0$, and so, the result holds.

Assume $0 < x < 1$. Since $f(t) = \sin t$ is differentiable for all real numbers, in particular, it is continuous on $[0, x]$ and differentiable on $(0, x)$. So, by the Mean Value Theorem (Theorem 10.25), there is $c \in (0, x)$ such that $\frac{f(x)-f(0)}{x-0} = f'(c)$, or equivalently, $\frac{\sin x}{x} = \cos c$. Since $-1 \leq \cos x \leq 1$, it follows that $-1 \leq \frac{\sin x}{x} \leq 1$, or equivalently, $-x \leq \sin x \leq x$, as desired. $\qquad\square$

10. Suppose that $(a, b) \subseteq A \subseteq \mathbb{R}$ and $f: A \to \mathbb{R}$ is differentiable on (a, b). Prove each of the following:
   (i)   If $f'(x) \geq 0$ for all $x \in (a, b)$, then f is increasing on (a, b).
   (ii) If $f'(x) \leq 0$ for all $x \in (a, b)$, then f is decreasing on (a, b).

Proofs: Let $x, y \in (a, b)$ with $x < y$. Then f is differentiable on $[x, y]$. By Theorem 10.11, f is continuous on $[x, y]$. By the Mean Value Theorem (Theorem 10.25), there is $c \in (x, y)$ such that $f'(c) = \frac{f(y)-f(x)}{y-x}$. So, $f(y) - f(x) = f'(c)(y - x)$. Since $x < y$, $y - x > 0$.

   (i)   Since $f'(c) \geq 0$, we have $f(y) - f(x) \geq 0$, or equivalently, $f(x) \leq f(y)$. Since $x, y \in (a, b)$ with $x < y$ were arbitrary, f is increasing on (a, b).

   (ii) Since $f'(c) \leq 0$, then $f(y) - f(x) \leq 0$, or equivalently, $f(x) \geq f(y)$. Since $x, y \in (a, b)$ with $x < y$ were arbitrary, f is decreasing on (a, b). $\qquad\square$

11. Prove the following special case of infinite version A of **L'Hôpital's Rule:** Suppose that $(a, c) \cup (c, b) \subseteq A \subseteq \mathbb{R}$, $f, g: A \to \mathbb{R}$ are continuous on A and differentiable on $(a, c) \cup (c, b)$, $\lim\limits_{x \to c} \frac{f(x)}{g(x)} = L$, where L is a finite nonzero real number, $\lim\limits_{x \to c} f(x) = \lim\limits_{x \to c} g(x) = \infty$ (either or both can also be $-\infty$), and $g'(x) \neq 0$ for all $x \in (a, c) \cup (c, b)$. If $\lim\limits_{x \to c} \frac{f'(x)}{g'(x)}$ exists, then

$$\lim_{x \to c} \frac{f(x)}{g(x)} = \lim_{x \to c} \frac{f'(x)}{g'(x)}.$$

Proof: Since $\lim\limits_{x \to c} f(x) = \lim\limits_{x \to c} g(x) = \infty$, we have $\lim\limits_{x \to c} \frac{1}{f(x)} = \lim\limits_{x \to c} \frac{1}{g(x)} = 0$ (Check this!). By standard version C of L'Hôpital's rule (Theorem 10.31), we have

$$L = \lim_{x \to c} \frac{f(x)}{g(x)} = \lim_{x \to c} \frac{\frac{1}{g(x)}}{\frac{1}{f(x)}} = \lim_{x \to c} \frac{\frac{d}{dx}\left[\frac{1}{g(x)}\right]}{\frac{d}{dx}\left[\frac{1}{f(x)}\right]} = \lim_{x \to c} \frac{\frac{-g'(x)}{[g(x)]^2}}{\frac{-f'(x)}{[f(x)]^2}} = \lim_{x \to c} \frac{[f(x)]^2 g'(x)}{[g(x)]^2 f'(x)}$$

$$= \lim_{x \to c} \frac{[f(x)]^2}{[g(x)]^2} \cdot \lim_{x \to c} \frac{g'(x)}{f'(x)} = \left[\lim_{x \to c} \frac{f(x)}{g(x)}\right]^2 \cdot \frac{1}{\lim\limits_{x \to c} \frac{f'(x)}{g'(x)}} = \frac{L^2}{\lim\limits_{x \to c} \frac{f'(x)}{g'(x)}}.$$

It follows that $L \cdot \lim\limits_{x \to c} \frac{f'(x)}{g'(x)} = L^2$, or equivalently, $L = \lim\limits_{x \to c} \frac{f'(x)}{g'(x)}$, as desired. □

LEVEL 4

12. Suppose that $[a, b] \subseteq A \subseteq \mathbb{R}$ and $f, g: A \to \mathbb{R}$ are functions that are continuous on $[a, b]$ and differentiable on (a, b). Prove that there is $c \in (a, b)$ such that

$$[f(b) - f(a)]g'(c) = [g(b) - g(a)]f'(c).$$

This is the **Generalized Mean Value Theorem** (Theorem 10.27).

Proof: Define $h: A \to \mathbb{R}$ by $h(x) = [f(b) - f(a)]g(x) - [g(b) - g(a)]f(x)$. Then h is continuous on $[a, b]$ and differentiable on (a, b) (Check this!).

Also, note that

$$h(a) = [f(b) - f(a)]g(a) - [g(b) - g(a)]f(a) = f(b)g(a) - f(a)g(a) - g(b)f(a) + g(a)f(a)$$
$$= f(b)g(a) - g(b)f(a),$$

$$h(b) = [f(b) - f(a)]g(b) - [g(b) - g(a)]f(b) = f(b)g(b) - f(a)g(b) - g(b)f(b) + g(a)f(b)$$
$$= g(a)f(b) - f(a)g(b) = f(b)g(a) - g(b)f(a).$$

So, $h(a) = h(b)$.

By Rolles's Theorem (Theorem 10.24), there is $c \in (a, b)$ such that $h'(c) = 0$. This is equivalent to $[f(b) - f(a)]g'(c) - [g(b) - g(a)]f'(c) = 0$ or $[f(b) - f(a)]g'(c) = [g(b) - g(a)]f'(c)$. □

13. Suppose that $f: (a, b) \to \mathbb{R}$ is differentiable on (a, b) with $f'(x) > 0$ for all $x \in (a, b)$. Prove that f is invertible. Let $g: B \to \mathbb{R}$ be the inverse of f. Prove that g is differentiable on B and $g'(f(x)) = \frac{1}{f'(x)}$ for all $x \in (a, b)$.

Proof: Since $f'(x) > 0$ for all $x \in (a, b)$, by Corollary 10.26, f is strictly increasing on (a, b). Suppose $x, y \in (a, b)$ with $x \neq y$. Without loss of generality, we may assume that $x < y$. Since f is strictly increasing, $f(x) < f(y)$. In particular, $f(x) \neq f(y)$. This shows that f is injective. By the Intermediate Value Theorem (Corollary 9.9) and the fact that f is strictly increasing, the range of f is the open interval $\left(f(a), f(b)\right)$. Therefore, $f: (a, b) \to \left(f(a), f(b)\right)$ is a bijection, and so, f is invertible.

Let $g: \left(f(a), f(b)\right) \to (a, b)$ be the inverse of f. It follows from Problem 9 from Problem Set 9 that g is continuous (Check this carefully!). Let $x, x + h \in (a, b)$, let $y = f(x)$ and $y + k = f(x + h)$. Then we have

$$\frac{g(y + k) - g(y)}{k} = \frac{(x + h) - x}{f(x + h) - y} = \frac{h}{f(x + h) - f(x)} = \frac{1}{\dfrac{f(x + h) - f(x)}{h}}.$$

Since $f'(x) \neq 0$ for all $x \in (a, b)$, it follows that for each $x \in (a, b)$ with $f(x) = y$,

$$\frac{1}{f'(x)} = \frac{1}{\displaystyle\lim_{h \to 0} \frac{f(x + h) - f(x)}{h}} = \lim_{h \to 0} \frac{1}{\dfrac{f(x + h) - f(x)}{h}} = \lim_{k \to 0} \frac{g(y + k) - g(y)}{k} = g'(y) = g'(f(x)).$$

$\square$

Note: In the last sequence of equalities above, we have

$$\lim_{h \to 0} \frac{1}{\dfrac{f(x + h) - f(x)}{h}} = \lim_{k \to 0} \frac{g(y + k) - g(y)}{k}.$$

In particular, we used the fact that for fixed $x \in (a, b)$, $h \to 0$ if and only if $k \to 0$. To see that this is true, first observe that $k = f(x + h) - f(x)$. Since f is continuous, as $h \to 0$, $k \to f(x) - f(x) = 0$. Next, observe that $h = g(y + k) - g(y)$. Since g is continuous, as $k \to 0$, $h \to g(x) - g(x) = 0$.

---

14. A function $f: A \to \mathbb{R}$ that is differentiable on A is said to be **uniformly differentiable** on A if

$$\forall \epsilon > 0 \; \exists \delta > 0 \; \forall x, y \in A \left(0 < |y - x| < \delta \to \left|\frac{f(y) - f(x)}{y - x} - f'(x)\right| < \epsilon\right).$$

Let $g: [a, b] \to \mathbb{R}$ be differentiable on $[a, b]$ with g' continuous on $[a, b]$. Prove that g is uniformly differentiable on $[a, b]$.

Proof: Since g' is continuous on $[a, b]$, by Problem 10 from Problem Set 9, g' is uniformly continuous on $[a, b]$. Let $\epsilon > 0$ be given. Since g' is uniformly continuous on $[a, b]$, there is $\delta > 0$ such that for all $u, v \in [a, b]$, $|v - u| < \delta \to |g'(v) - g'(u)| < \epsilon$.

Suppose that $x, y \in [a, b]$ and $0 < |y - x| < \delta$. By the Mean Value Theorem, there is $c \in (a, b)$ such that $\frac{g(y) - g(x)}{y - x} = g'(c)$. It follows that

$$\left|\frac{g(y) - g(x)}{y - x} - g'(x)\right| = |g'(c) - g'(x)| < \epsilon.$$

Since $\epsilon > 0$ was arbitrary, g is uniformly differentiable on $[a, b]$. $\square$

15. Provide a counterexample to show that the following statement is false: If $f: \mathbb{R} \to \mathbb{R}$ satisfies $f(0) = 0$ and $f'(0) = 1$, then there is a neighborhood N of 0 such that f is strictly monotonic on N.

Solution: Define $f: \mathbb{R} \to \mathbb{R}$ by $f(x) = \begin{cases} 2x^2 \sin\frac{1}{x} + x & \text{if } x \neq 0 \\ 0 & \text{if } x = 0 \end{cases}$. By definition, $f(0) = 0$. Also, we have

$$f'(0) = \lim_{x \to 0} \frac{f(x) - f(0)}{x - 0} = \lim_{x \to 0} \left(2x \sin\frac{1}{x} + 1\right) = 0 + 1 = 1.$$

(Use the Squeeze Theorem (Problem 14 from Problem Set 8) to show that $\lim_{x \to 0} 2x \sin\frac{1}{x} = 0$.)

Now, if $x \neq 0$, then $f'(x) = 2x^2 \left(\cos\frac{1}{x}\right)\left(-\frac{1}{x^2}\right) + 4x \sin\frac{1}{x} + 1 = -2\cos\frac{1}{x} + 4x \sin\frac{1}{x} + 1$.

$$f'\left(\frac{1}{2n\pi}\right) = -2\cos 2n\pi + \frac{2}{n\pi}\sin 2n\pi + 1 = -2 + 0 + 1 = -1 < 0.$$

$$f'\left(\frac{2}{(4n+1)\pi}\right) = -2\cos\frac{(4n+1)\pi}{2} + \frac{8}{(4n+1)\pi}\sin\frac{(4n+1)\pi}{2} + 1 = 0 + \frac{8}{(4n+1)\pi} + 1 > 0.$$

Since the sequences $\left(\frac{1}{2n\pi}\right)$ and $\left(\frac{2}{(4n+1)\pi}\right)$ both converge to 0, we see that f' takes on both positive and negative values in every neighborhood of 0. Therefore, f is **not** strictly monotonic on any neighborhood of 0.

16. Suppose that $f: (0, \infty) \to \mathbb{R}$ is differentiable on $(0, \infty)$, $f'(x)$ is uniformly continuous on $(0, \infty)$, and $\lim_{x \to \infty} f(x)$ is a finite real number. Prove that $\lim_{x \to \infty} f'(x) = 0$.

Proof: Assume toward contradiction that $\lim_{x \to \infty} f'(x) \neq 0$. Without loss of generality, we may assume that $\lim_{x \to \infty} f'(x) \not< 0$ (if $\lim_{x \to \infty} f'(x) < 0$, simply replace f by $-f$). Then there is $\epsilon > 0$ and a sequence (x_n) such that $x_n \to \infty$ and $f'(x_n) \geq \epsilon > 0$. Since $f'(x)$ is uniformly continuous on $(0, \infty)$, we can choose $\delta > 0$ so that

$$\forall x, y \in (0, \infty) \left(|y - x| < \delta \to |f'(y) - f'(x)| < \frac{\epsilon}{2}\right).$$

Then we have

$$|y - x_n| < \delta \to |f'(y) - f'(x_n)| < \frac{\epsilon}{2} \to -\frac{\epsilon}{2} < f'(y) - f'(x_n) < \frac{\epsilon}{2}$$

$$\to f'(y) > f'(x_n) - \frac{\epsilon}{2} \to f'(y) > \epsilon - \frac{\epsilon}{2} \to f'(y) > \frac{\epsilon}{2}.$$

Now, $|y - x_n| < \delta$ if and only if $-\delta < y - x_n < \delta$ if and only if $x_n - \delta < y < x_n + \delta$ if and only if $y \in (x_n - \delta, x_n + \delta)$.

By the Mean Value Theorem, there is a c between $x_n - \delta$ and $x_n + \delta$ such that

$$\frac{f(x_n + \delta) - f(x_n - \delta)}{(x_n + \delta) - (x_n - \delta)} = f'(c)$$

So,

$$\frac{|f(x_n + \delta) - f(x_n - \delta)|}{2\delta} = \left| \frac{f(x_n + \delta) - f(x_n - \delta)}{2\delta} \right| = \left| \frac{f(x_n + \delta) - f(x_n - \delta)}{(x_n + \delta) - (x_n - \delta)} \right| = |f'(c)| > \frac{\epsilon}{2}.$$

It follows that $|f(x_n + \delta) - f(x_n - \delta)| > \frac{\epsilon}{2} \cdot 2\delta = \epsilon\delta$.

Since $\epsilon\delta > 0$, there is $K \in \mathbb{N}$ such that $x > K \rightarrow |f(x) - L| < \frac{\epsilon\delta}{2}$, where $\lim\limits_{x \to \infty} f(x) = L$.

Therefore, by the Triangle Inequality and the fact that $x_n \to \infty$, for n sufficiently large, we have

$$|f(x_n + \delta) - f(x_n - \delta)| \leq |f(x_n + \delta) - L| + |L - f(x_n - \delta)| < \frac{\epsilon\delta}{2} + \frac{\epsilon\delta}{2} = \epsilon\delta.$$

We have now reached a contradiction. $\qquad\qquad\qquad\qquad\qquad\qquad\qquad\qquad\quad$ $\square$

Problem Set 11

LEVEL 1

1. Prove each of the following:

   (i) For each $n \in \mathbb{Z}^+$,
   $$\sum_{i=1}^{n} 1 = n.$$

   (ii) Let $n \in \mathbb{Z}^+$ and let $r, a_1, a_2, \ldots, a_n \in \mathbb{R}$. Then
   $$\sum_{i=1}^{n} ra_i = r \sum_{i=1}^{n} a_i.$$

   (iii) Let $n \in \mathbb{Z}^+$. For each $i = 1, 2, \ldots, n$, let $a_i, b_i \in \mathbb{R}$. Then
   $$\sum_{i=1}^{n} (a_i + b_i) = \sum_{i=1}^{n} a_i + \sum_{i=1}^{n} b_i.$$

Proofs:

(i) **Base Case** ($k = 1$):
$$\sum_{i=1}^{1} 1 = 1.$$

Inductive Step: Let $k \in \mathbb{N}$ with $k \geq 1$ and assume that
$$\sum_{i=1}^{k} 1 = k.$$

Then we have
$$\sum_{i=1}^{k+1} 1 = \left(\sum_{i=1}^{k} 1 \right) + 1 = k + 1.$$

By the Principle of Mathematical Induction, the result holds for all natural numbers $n \geq 1$. $\quad\square$

(ii) Let $r \in \mathbb{R}$.

Base Case ($k = 1$):
$$\sum_{i=1}^{1} ra_i = ra_i = r \sum_{i=1}^{1} a_i.$$

Inductive Step: Let $k \in \mathbb{N}$ with $k \geq 1$ and assume that

$$\sum_{i=1}^{k} ra_i = r \sum_{i=1}^{k} a_i.$$

Then we have

$$\sum_{i=1}^{k+1} ra_i = \left(\sum_{i=1}^{k} ra_i \right) + ra_{k+1} = \left(r \sum_{i=1}^{k} a_i \right) + ra_{k+1} = r \left(\left(\sum_{i=1}^{k} a_i \right) + a_{k+1} \right) = r \sum_{i=1}^{k+1} a_i.$$

By the Principle of Mathematical Induction, the result holds for all natural numbers $n \geq 1$. □

(iii) For each $i = 1, 2, \ldots, n$, let $a_i, b_i \in \mathbb{R}$.

Base Case ($k = 1$):

$$\sum_{i=1}^{1} (a_i + b_i) = a_1 + b_1 = \sum_{i=1}^{1} a_i + \sum_{i=1}^{1} b_i.$$

Inductive Step: Let $k \in \mathbb{N}$ with $k \geq 1$ and assume that

$$\sum_{i=1}^{k} (a_i + b_i) = \sum_{i=1}^{k} a_i + \sum_{i=1}^{k} b_i.$$

Then we have

$$\sum_{i=1}^{k+1} (a_i + b_i) = \left(\sum_{i=1}^{k} (a_i + b_i) \right) + (a_{k+1} + b_{k+1}) = \left(\sum_{i=1}^{k} a_i + \sum_{i=1}^{k} b_i \right) + (a_{k+1} + b_{k+1})$$

$$= \left[\left(\sum_{i=1}^{k} a_i \right) + a_{k+1} \right] + \left[\left(\sum_{i=1}^{k} b_i \right) + b_{k+1} \right] = \sum_{i=1}^{k+1} a_i + \sum_{i=1}^{k+1} b_i.$$

By the Principle of Mathematical Induction, the result holds for all natural numbers $n \geq 1$. □

2. Let $f : [0, 3] \to \mathbb{R}$ be defined by $f(x) = x$.

   (i) Let $P = \{[0, 1], [1, 2], [2, 3]\}$. Compute $L_f(P)$ and $U_f(P)$.

   (ii) Let $n \in \mathbb{Z}^+$ and let P_n be the partition of $[0, 3]$ into n equal subintervals. Compute $L_f(P_n)$ and $U_f(P_n)$.

   (iii) Compute $\underline{\int_a^b} f(x)\, dx$ and $\overline{\int_a^b} f(x)\, dx$. (You may use Problem 18 below.)

   (iv) Is f Riemann integrable on $[0, 3]$? If so, compute $\int_0^3 f(x)\, dx$. If not, explain why.

Solutions:

   (i) $L_f(P) = 0 \cdot 1 + 1 \cdot 1 + 2 \cdot 1 = 0 + 1 + 2 = 3$; $U_f(P) = 1 \cdot 1 + 2 \cdot 1 + 3 \cdot 1 = 1 + 2 + 3 = 6$

   (ii) $\Delta x = \frac{3-0}{n} = \frac{3}{n}$ and $P_n = \left\{ \left[\frac{3(i-1)}{n}, \frac{3i}{n} \right] \right\}_{i=1}^{n}$. So, we have

$$L_f(P) = \sum_{i=1}^{n} m_i \Delta x = \sum_{i=1}^{n} \left(\frac{3(i-1)}{n}\right)\frac{3}{n} = \frac{9}{n^2}\sum_{i=1}^{n}(i-1) = \frac{9}{n^2}\left(\sum_{i=1}^{n} i - \sum_{i=1}^{n} 1\right)$$

$$= \frac{9}{n^2}\left(\frac{n(n+1)}{2} - n\right) = \frac{9(n+1)}{2n} - \frac{9}{n} = \frac{9n+9}{2n} - \frac{18}{2n} = \frac{9n-9}{2n}$$

$$U_f(P) = \sum_{i=1}^{n} M_i \Delta x = \sum_{i=1}^{n} \frac{3i}{n}\cdot\frac{3}{n} = \frac{9}{n^2}\sum_{i=1}^{n} i = \frac{9}{n^2}\cdot\frac{n(n+1)}{2} = \frac{9(n+1)}{2n} = \frac{9n+9}{2n}$$

(iii) Using Problem 18 below, we have

$$\underline{\int_0^3} f(x) = \sup\left\{\frac{9n-9}{2n} \;\middle|\; n \in \mathbb{Z}^+\right\} = \lim_{n\to\infty}\frac{9n-9}{2n} = \frac{9}{2}$$

$$\overline{\int_0^3} f(x)\,dx = \inf\left\{\frac{9n+9}{2n} \;\middle|\; n \in \mathbb{Z}^+\right\} = \lim_{n\to\infty}\frac{9n+9}{2n} = \frac{9}{2}$$

(iv) Since $\overline{\int_0^3} f(x)\,dx = \underline{\int_0^3} f(x)$, f is Riemann integrable on $[0,3]$ and

$$\int_0^3 f(x)\,dx = \frac{9}{2}.$$

LEVEL 2

3. Let A and B be nonempty bounded sets of real numbers with $A \subseteq B$. Prove that
$$\inf B \le \inf A \le \sup A \le \sup B.$$

Proof: Let $k = \inf B$. Then k is a lower bound of B. So, for all $b \in B$, $k \le b$. Since $A \subseteq B$, we have that for all $b \in A$, $k \le b$. So, k is a lower bound of A. Since $\inf A$ is the **greatest** lower bound of A, it follows that $k \le \inf A$. Since $k = \inf B$, we have $\inf B \le \inf A$.

Now, let $k = \inf A$. Since $A \ne \emptyset$, there is $a \in A$ with $k \le a$. Since $\sup A$ is an upper bound of A, we have $a \le \sup A$. So, $\inf A = k \le a \le \sup A$.

Finally, let $k = \sup B$. Then k is an upper bound of B. So, for all $b \in B$, $b \le k$. Since $A \subseteq B$, we have that for all $b \in A$, $b \le k$. So, k is an upper bound of A. Since $\sup A$ is the **least** upper bound of A, it follows that $\sup A \le k$. Since $k = \sup B$, we have $\sup A \le \sup B$. □

4. Let A be a nonempty bounded set of real numbers, let $c \ge 0$, and let $B = \{cx \mid x \in A\}$. Prove that $\sup B = c \sup A$ and $\inf B = c \inf A$.

Proof: If $c = 0$, then $B = \{0\}$, and so, $\sup B = 0$, $\inf B = 0$, $c \sup A = 0$, and $c \inf A = 0$. So, we have $\sup B = c \sup A$ and $\inf B = c \inf A$.

Now, assume $c > 0$.

Let $cx \in B$. Then $x \in A$. Since $\sup A$ is an upper bound of A, $x \leq \sup A$. Since $c > 0$, we have $cx \leq c \sup A$. Since $cx \in B$ was arbitrary, $c \sup A$ is an upper bound of B. Since $\sup B$ is the **least** upper bound of B, $\sup B \leq c \sup A$.

Let $x \in A$. Then $cx \in B$. Since $\sup B$ is an upper bound of B, we have $cx \leq \sup B$. Since $c > 0$, it follows that $x \leq \frac{1}{c} \sup B$. Since $x \in A$ was arbitrary, $\frac{1}{c} \sup B$ is an upper bound of A. Since $\sup A$ is the **least** upper bound of A, $\sup A \leq \frac{1}{c} \sup B$, and therefore, $c \sup A \leq \sup B$.

Since $\sup B \leq c \sup A$ and $c \sup A \leq \sup B$, it follows that $\sup B = c \sup A$.

Now, let $cx \in B$. Then $x \in A$. Since $\inf A$ is a lower bound of A, $x \geq \inf A$. Since $c > 0$, we have $cx \geq c \inf A$. Since $cx \in B$ was arbitrary, $c \inf A$ is a lower bound of B. Since $\inf B$ is the **greatest** lower bound of B, $\inf B \geq c \inf A$.

Let $x \in A$. Then $cx \in B$. Since $\inf B$ is a lower bound of B, we have $cx \geq \inf B$. Since $c > 0$, it follows that $x \geq \frac{1}{c} \inf B$. Since $x \in A$ was arbitrary, $\frac{1}{c} \inf B$ is a lower bound of A. Since $\inf A$ is the **greatest** lower bound of A, $\inf A \geq \frac{1}{c} \inf B$, and therefore, $c \inf A \geq \inf B$.

Since $\inf B \geq c \inf A$ and $c \inf A \geq \inf B$, it follows that $\inf B = c \inf A$. $\qquad\square$

5. Let $f, g \colon [a, b] \to \mathbb{R}$ be bounded functions, let $A = \{f(x) \mid x \in [a, b]\}$, $B = \{g(x) \mid x \in [a, b]\}$, and $C = \{f(x) + g(x) \mid x \in [a, b]\}$ Prove that

$$\sup C \leq \sup A + \sup B \quad \text{and} \quad \inf C \geq \inf A + \inf B.$$

Proof: Let $x \in [a, b]$. Then $f(x) \in A$, and therefore, $\inf A \leq f(x) \leq \sup A$. Also, $g(x) \in B$, and therefore, $\inf B \leq g(x) \leq \sup B$. So, $\inf A + \inf B \leq f(x) + g(x) \leq \sup A + \sup B$. Since $x \in [a, b]$ was arbitrary, $\sup A + \sup B$ is an upper bound of C and $\inf A + \inf B$ is a lower bound of C. Since $\sup C$ is the **least** upper bound of C, $\sup C \leq \sup A + \sup B$. Similarly, since $\inf C$ is the **greatest** lower bound of C, $\inf C \geq \inf A + \inf B$. $\qquad\square$

6. Let $f, g \colon [a, b] \to \mathbb{R}$ be bounded functions that are Riemann integrable on $[a, b]$ and let $f(x) \leq g(x)$ for all $x \in [a, b]$. Prove that

$$\int_a^b f(x)\, dx \leq \int_a^b g(x)\, dx.$$

Proof: Let $P = \{[x_{i-1}, x_i]\}_{i=1}^n$ be a partition of $[a, b]$ and let $m_i = \inf\{f(x) \mid x_{i-1} \leq x \leq x_i\}$ and $k_i = \inf\{g(x) \mid x_{i-1} \leq x \leq x_i\}$ for $i = 1, 2, \ldots, n$. Then $m_i \leq k_i$ for each $i = 1, 2, \ldots, n$. So, we have

$$L_f(P) = \sum_{i=1}^n m_i \Delta x_i \leq \sum_{i=1}^n k_i \Delta x_i = L_g(P)$$

Since P was an arbitrary partition of $[a, b]$, we have

$$\underline{\int_a^b} f(x)\,dx = \sup\{L_f(P) \mid P \text{ a partition of } [a,b]\}$$

$$\leq \sup\{L_g(P) \mid P \text{ a partition of } [a,b]\} = \underline{\int_a^b} g(x)\,dx.$$

Since f and g are Riemann integrable on $[a,b]$, we have

$$\int_a^b f(x)\,dx = \underline{\int_a^b} f(x)\,dx \leq \underline{\int_a^b} g(x)\,dx = \int_a^b g(x)\,dx.$$

$\square$

LEVEL 3

7. Let A and B be nonempty bounded sets of real numbers and let $A + B = \{a + b \mid a \in A, b \in B\}$. Prove that $\sup(A+B) = \sup A + \sup B$ and $\inf(A+B) = \inf A + \inf B$.

Proof: Let $a + b \in A + B$, where $a \in A$ and $b \in B$. Since $\sup A$ is an upper bound of A, $a \leq \sup A$. Similarly, $b \leq \sup B$. So, $a + b \leq \sup A + \sup B$. Since $a + b \in A + B$ was arbitrary, it follows that $\sup A + \sup B$ is an upper bound of $A + B$. Since $\sup(A+B)$ is the **least** upper bound of $A + B$, we have $\sup(A+B) \leq \sup A + \sup B$.

Let $a \in A$. Then for each $b \in B$, we have $a + b \in A + B$. Since $\sup(A+B)$ is an upper bound of $A + B$, for each $b \in B$, $a + b \leq \sup(A+B)$. So, for each $b \in B$, $b \leq \sup(A+B) - a$. Therefore, $\sup(A+B) - a$ is an upper bound of B. Since $\sup B$ is the **least** upper bound of B, it follows that $\sup B \leq \sup(A+B) - a$. Therefore, $a \leq \sup(A+B) - \sup B$. Since $a \in A$ was arbitrary, we see that $\sup(A+B) - \sup B$ is an upper bound of A. Since $\sup A$ is the **least** upper bound of A, we have $\sup A \leq \sup(A+B) - \sup B$. Therefore, $\sup A + \sup B \leq \sup(A+B)$.

Since $\sup(A+B) \leq \sup A + \sup B$ and $\sup A + \sup B \leq \sup(A+B)$, $\sup(A+B) = \sup A + \sup B$.

Let $a + b \in A + B$, where $a \in A$ and $b \in B$. Since $\inf A$ is a lower bound of A, $a \geq \inf A$. Similarly, $b \geq \inf B$. So, $a + b \geq \inf A + \inf B$. Since $a + b \in A + B$ was arbitrary, it follows that $\inf A + \inf B$ is a lower bound of $A + B$. Since $\inf(A+B)$ is the **greatest** lower bound of $A + B$, it follows that $\inf(A+B) \geq \inf A + \inf B$.

Let $a \in A$. Then for each $b \in B$, we have $a + b \in A + B$. Since $\inf(A+B)$ is a lower bound of $A + B$, for each $b \in B$, $a + b \geq \inf(A+B)$. So, for each $b \in B$, $b \geq \inf(A+B) - a$. So, $\inf(A+B) - a$ is a lower bound of B. Since $\inf B$ is the **greatest** lower bound of B, it follows that $\inf B \geq \inf(A+B) - a$. Therefore, $a \geq \inf(A+B) - \inf B$. Since $a \in A$ was arbitrary, we see that $\inf(A+B) - \inf B$ is a lower bound of A. Since $\inf A$ is the **greatest** lower bound of A, we have $\inf A \geq \inf(A+B) - \inf B$. Therefore, $\inf A + \inf B \geq \inf(A+B)$.

Since $\inf(A+B) \geq \inf A + \inf B$ and $\inf A + \inf B \geq \inf(A+B)$, $\inf(A+B) = \inf A + \inf B$. $\square$

8. Let $f: [a, b] \to \mathbb{R}$ be continuous on $[a, b]$. Prove that there is $c \in [a, b]$ such that

$$\int_a^b f(x)\, dx = f(c)(b - a).$$

This result is known as the **Mean Value Theorem for Integrals**.

Proof: Define $F: [a, b] \to \mathbb{R}$ by $F(x) = \int_a^x f(t)\, dt$. By the first form of the Fundamental Theorem of Calculus (Theorem 11.12), F is continuous on $[a, b]$ and differentiable on (a, b). Therefore, by the Mean Value Theorem (Theorem 10.25), there is $c \in (a, b)$ such that

$$F'(c) = \frac{F(b) - F(a)}{b - a} = \frac{\int_a^b f(t)\, dt - \int_a^a f(t)\, dt}{b - a} = \frac{\int_a^b f(t)\, dt - 0}{b - a} = \frac{1}{b - a} \int_a^b f(t)\, dt.$$

Again, by the first form of the Fundamental Theorem of Calculus, we have $F'(c) = f(c)$. Therefore,

$$\frac{1}{b - a} \int_a^b f(t)\, dt = f(c).$$

Equivalently, we have

$$\int_a^b f(x)\, dx = f(c)(b - a).$$

$\square$

9. Let $f: [a, b] \to \mathbb{R}$ be a bounded function that is Riemann integrable on $[a, b]$ and let $c \in \mathbb{R}$. Prove that cf is Riemann integrable on $[a, b]$ and

$$\int_a^b cf(x)\, dx = c \int_a^b f(x)\, dx.$$

Proof: First assume that $c \geq 0$. Let $P = \{[x_{i-1}, x_i]\}_{i=1}^n$ be a partition of $[a, b]$ and define m_i and M_i for $i = 1, 2, \ldots, n$ by $m_i = \inf\{f(x) \mid x_{i-1} \leq x \leq x_i\}$ and $M_i = \sup\{f(x) \mid x_{i-1} \leq x \leq x_i\}$. By Problem 4 above, $cm_i = \inf\{cf(x) \mid x_{i-1} \leq x \leq x_i\}$ and $cM_i = \sup\{cf(x) \mid x_{i-1} \leq x \leq x_i\}$. So, we have the following:

$$L_{cf}(P) = \sum_{i=1}^n cm_i \Delta x_i = c \sum_{i=1}^n m_i \Delta x_i = cL_f(P)$$

$$U_{cf}(P) = \sum_{i=1}^n cM_i \Delta x_i = c \sum_{i=1}^n M_i \Delta x_i = cU_f(P)$$

By Problem 4 again, we have the following:

$$\underline{\int_a^b} cf(x)\, dx = \sup\{L_{cf}(P) \mid P \text{ a partition of } [a, b]\} = \sup\{cL_f(P) \mid P \text{ a partition of } [a, b]\}$$

$$= c \sup\{L_f(P) \mid P \text{ a partition of } [a, b]\} = c \underline{\int_a^b} f(x)\, dx$$

$$\overline{\int_a^b} cf(x)\,dx = \inf\{U_{cf}(P) \mid P \text{ a partition of } [a,b]\} = \inf\{cU_f(P) \mid P \text{ a partition of } [a,b]\}$$

$$= c\inf\{U_f(P) \mid P \text{ a partition of } [a,b]\} = c\overline{\int_a^b} f(x)\,dx$$

Since f is Riemann integrable on $[a,b]$, we have

$$\underline{\int_a^b} cf(x)\,dx = c\underline{\int_a^b} f(x)\,dx = c\overline{\int_a^b} f(x)\,dx = \overline{\int_a^b} cf(x)\,dx.$$

It follows that cf is Riemann integrable on $[a,b]$ and $\int_a^b cf(x)\,dx = c\int_a^b f(x)\,dx$.

Next, assume that $c = -1$ and P, m_i, and M_i are defined the same way as above. By the method used in the proof of Problem 18 in Problem Set 6, we see that $-m_i = \sup\{-f(x) \mid x_{i-1} \le x \le x_i\} = K_i$ and $-M_i = \inf\{-f(x) \mid x_{i-1} \le x \le x_i\} = k_i$ (k_i and K_i are defined as shown). So, we have the following:

$$L_{-f}(P) = \sum_{i=1}^n k_i \Delta x_i = \sum_{i=1}^n -M_i \Delta x_i = -\sum_{i=1}^n M_i \Delta x_i = -U_f(P)$$

$$U_{-f}(P) = \sum_{i=1}^n K_i \Delta x_i = \sum_{i=1}^n -m_i \Delta x_i = -\sum_{i=1}^n m_i \Delta x_i = -L_f(P)$$

By the same argument as above, we have the following:

$$\underline{\int_a^b} -f(x)\,dx = \sup\{L_{-f}(P) \mid P \text{ a partition of } [a,b]\} = \sup\{-U_f(P) \mid P \text{ a partition of } [a,b]\}$$

$$= -\inf\{U_f(P) \mid P \text{ a partition of } [a,b]\} = -\overline{\int_a^b} f(x)\,dx$$

$$\overline{\int_a^b} -f(x)\,dx = \inf\{U_{-f}(P) \mid P \text{ a partition of } [a,b]\} = \inf\{-L_f(P) \mid P \text{ a partition of } [a,b]\}$$

$$= -\sup\{L_f(P) \mid P \text{ a partition of } [a,b]\} = -\underline{\int_a^b} f(x)\,dx$$

Since f is Riemann integrable on $[a,b]$, we have

$$\underline{\int_a^b} -f(x)\,dx = -\overline{\int_a^b} f(x)\,dx = -\underline{\int_a^b} f(x)\,dx = \overline{\int_a^b} -f(x)\,dx.$$

So, $-f$ is Riemann integrable on $[a,b]$ and $\int_a^b -f(x)\,dx = -\int_a^b f(x)\,dx$.

Finally, if $c < 0$, then we have the following:

$$\int_{\underline{a}}^{b} cf(x)\,dx = \int_{\underline{a}}^{b} -(-cf(x))\,dx = -\overline{\int_{a}^{b}} -cf(x)\,dx = -(-c)\overline{\int_{a}^{b}} f(x)\,dx = c\overline{\int_{a}^{b}} f(x)\,dx$$

$$\overline{\int_{a}^{b}} cf(x)\,dx = \overline{\int_{a}^{b}} -(-cf(x))\,dx = -\int_{\underline{a}}^{b} -cf(x)\,dx = -(-c)\int_{\underline{a}}^{b} f(x)\,dx = c\int_{\underline{a}}^{b} f(x)\,dx$$

Since f is Riemann integrable on $[a, b]$, we have

$$\int_{\underline{a}}^{b} cf(x)\,dx = c\overline{\int_{a}^{b}} f(x)\,dx = c\int_{\underline{a}}^{b} f(x)\,dx = \overline{\int_{a}^{b}} cf(x)\,dx.$$

It follows that cf is Riemann integrable and $\int_a^b cf(x)\,dx = c\int_a^b f(x)\,dx$. $\qquad\square$

LEVEL 4

10. Let $f:[a, b] \to \mathbb{R}$ be a bounded function and let P and Q be partitions of $[a,b]$ with Q a refinement of P. Prove that $L_f(P) \le L_f(Q)$ and $U_f(Q) \le U_f(P)$.

Proof: Let $P = \{[x_{i-1}, x_i]\}_{i=1}^{n}$ and $Q = \{[y_{i-1}, y_i]\}_{i=1}^{t}$. Since Q is a refinement of P, it follows that $\{x_0, x_1, \ldots, x_n\} \subseteq \{y_0, y_1, \ldots, y_t\}$. So, there are integers $k_0 < k_1 < \cdots < k_n$ with $k_0 = 0$ and $k_n = t$ and $x_i = y_{k_i}$ for each $i = 1, 2, \ldots, n$. We then have for each $i = 1, 2, \ldots, n$,

$$\Delta x_i = x_i - x_{i-1} = y_{k_i} - y_{k_{i-1}} = \sum_{j=k_{i-1}+1}^{k_i} y_j - y_{j-1} = \sum_{j=k_{i-1}+1}^{k_i} \Delta y_j.$$

For each $i = 1, 2, \ldots, n$, let $m_i = \inf\{f(x) \mid x_{i-1} \le x \le x_i\}$ and for each $j = 1, 2, \ldots, t$, let $n_j = \inf\{f(x) \mid y_{j-1} \le x \le y_j\}$. Then $m_i \le n_j$ for $k_{i-1} < j \le k_i$. So,

$$m_i \Delta x_i = m_i \sum_{j=k_{i-1}+1}^{k_i} \Delta y_j = \sum_{j=k_{i-1}+1}^{k_i} m_i \Delta y_j \le \sum_{j=k_{i-1}+1}^{k_i} n_j \Delta y_j.$$

Therefore, we have

$$L_f(P) = \sum_{i=1}^{n} m_i \Delta x_i \le \sum_{i=1}^{n} \sum_{j=k_{i-1}+1}^{k_i} n_j \Delta y_j = \sum_{i=1}^{t} n_i \Delta y_i = L_f(Q).$$

Similarly, for each $i = 1, 2, \ldots, n$, let $M_i = \sup\{f(x) \mid x_{i-1} \le x \le x_i\}$ and for each $j = 1, 2, \ldots, t$, let $N_j = \sup\{f(x) \mid y_{j-1} \le x \le y_j\}$. Then $M_i \ge N_j$ for $k_{i-1} < j \le k_i$. So,

$$M_i \Delta x_i = M_i \sum_{j=k_{i-1}+1}^{k_i} \Delta y_j = \sum_{j=k_{i-1}+1}^{k_i} M_i \Delta y_j \ge \sum_{j=k_{i-1}+1}^{k_i} N_j \Delta y_j.$$

Therefore, we have

$$U_f(P) = \sum_{i=1}^{n} M_i \Delta x_i \geq \sum_{i=1}^{n} \sum_{j=k_{i-1}+1}^{k_i} N_j \Delta y_j = \sum_{i=1}^{t} N_i \Delta y_i = U_f(Q).$$

□

11. Let $f:[a,b] \to \mathbb{R}$ be a bounded function, let $m = \inf\{f(x) \mid x \in [a,b]\}$, and let $M = \sup\{f(x) \mid x \in [a,b]\}$. Prove that

$$m(b-a) \leq \underline{\int_a^b} f(x)\, dx \leq \overline{\int_a^b} f(x)\, dx \leq M(b-a).$$

Proof: Let P be any partition of $[a,b]$. By Theorem 11.5, we have

$$m(b-a) \leq L_f(P) \leq U_f(P) \leq M(b-a).$$

Since $L_f(P) \leq \underline{\int_a^b} f(x)\, dx$, we have $m(b-a) \leq \underline{\int_a^b} f(x)\, dx$. Similarly, since $\overline{\int_a^b} f(x)\, dx \leq U_f(P)$, we have $\overline{\int_a^b} f(x)\, dx \leq M(b-a)$.

Now, let Q and R be partitions of $[a,b]$ and let $T = Q * R$ (see Note 2 before Example 11.6). By Problem 10, we have $L_f(Q) \leq L_f(T)$ and $U_f(T) \leq U_f(R)$. By Theorem 11.5, we have

$$L_f(Q) \leq L_f(T) \leq U_f(T) \leq U_f(R).$$

So, for any partitions Q and R of $[a,b]$, we have $L_f(Q) \leq U_f(R)$.

It follows that if Q is a fixed partition of $[a,b]$, then $L_f(Q)$ is a lower bound of $\{U_f(P) \mid P \text{ is a partition of } [a,b]\}$. Therefore, $L_f(Q) \leq \inf\{U_f(P) \mid P \text{ is a partition of } [a,b]\}$ (this is true because $\inf\{U_f(P) \mid P \text{ is a partition of } [a,b]\}$ is the **greatest** lower bound of $\{U_f(P) \mid P \text{ is a partition of } [a,b]\}$).

Since Q was an arbitrary partition of $[a,b]$, it follows that $\inf\{U_f(P) \mid P \text{ is a partition of } [a,b]\}$ is an upper bound of $\{L_f(P) \mid P \text{ is a partition of } [a,b]\}$. Since $\sup\{L_f(P) \mid P \text{ is a partition of } [a,b]\}$ is the **least** upper bound of $\{L_f(P) \mid P \text{ is a partition of } [a,b]\}$, it follows that $\sup\{L_f(P) \mid P \text{ is a partition of } [a,b]\} \leq \inf\{U_f(P) \mid P \text{ is a partition of } [a,b]\}$.

Therefore,

$$\underline{\int_a^b} f(x)\, dx = \sup\{L_f(P) \mid P \text{ is a partition of } [a,b]\}$$

$$\leq \inf\{U_f(P) \mid P \text{ is a partition of } [a,b]\} = \overline{\int_a^b} f(x)\, dx.$$

□

12. Let $f, g: [a, b] \to \mathbb{R}$ be bounded functions, let $A = \{L_f(P) \mid P \text{ is a partition of } [a, b]\}$, $B = \{L_g(P) \mid P \text{ is a partition of } [a, b]\}$, and $C = \{L_f(P) + L_g(P) \mid P \text{ is a partition of } [a, b]\}$. Prove that $\sup C = \sup A + \sup B$. Then prove an analogous result involving upper sums.

Proof: Let $L_f(P) + L_g(P) \in C$. Then $L_f(P) \in A$ and $L_g(P) \in B$. Since $\sup A$ is an upper bound of A, we have $L_f(P) \leq \sup A$. Similarly, $L_g(P) \leq \sup B$. So, $L_f(P) + L_g(P) \leq \sup A + \sup B$. Since $L_f(P) + L_g(P) \in C$ was arbitrary, it follows that $\sup A + \sup B$ is an upper bound of C. Since $\sup C$ is the **least** upper bound of C, we have $\sup C \leq \sup A + \sup B$.

Fix $L_f(P) \in A$. Now, let $L_g(Q) \in B$ and let $R = P * Q$ (see Note 2 before Example 11.6). By Problem 10 above, we have $L_f(P) \leq L_f(R)$ and $L_g(Q) \leq L_g(R)$. So, $L_f(P) + L_g(Q) \leq L_f(R) + L_g(R)$. Since $L_f(R) + L_g(R) \in C$, we see that $L_f(P) + L_g(Q) \leq L_f(R) + L_g(R) \leq \sup C$. Therefore, $L_g(Q) \leq \sup C - L_f(P)$. Since $L_g(Q) \in B$ was arbitrary, $\sup C - L_f(P)$ is an upper bound of B. Since $\sup B$ is the **least** upper bound of B, it follows that $\sup B \leq \sup C - L_f(P)$. Therefore, $L_f(P) \leq \sup C - \sup B$. Since $L_f(P) \in A$ was arbitrary, we see that $\sup C - \sup B$ is an upper bound of A. Since $\sup A$ is the **least** upper bound of A, we have $\sup A \leq \sup C - \sup B$. Therefore, $\sup A + \sup B \leq \sup C$.

Since $\sup C \leq \sup A + \sup B$ and $\sup A + \sup B \leq \sup C$, we have $\sup C = \sup A + \sup B$.

Next, we let $A = \{U_f(P) \mid P \text{ is a partition of } [a, b]\}$, $B = \{U_g(P) \mid P \text{ is a partition of } [a, b]\}$, and $C = \{U_f(P) + U_g(P) \mid P \text{ is a partition of } [a, b]\}$. We will prove that $\inf C = \inf A + \inf B$.

Let $U_f(P) + U_g(P) \in C$. Then $U_f(P) \in A$ and $U_g(P) \in B$. Since $\inf A$ is a lower bound of A, we have $U_f(P) \geq \inf A$. Similarly, $U_g(P) \geq \inf B$. So, $U_f(P) + U_g(P) \geq \inf A + \inf B$. Since $U_f(P) + U_g(P) \in C$ was arbitrary, it follows that $\inf A + \inf B$ is a lower bound of C. Since $\inf C$ is the **greatest** lower bound of C, we have $\inf C \geq \inf A + \inf B$.

Fix $U_f(P) \in A$. Now, let $U_g(Q) \in B$ and let $R = P * Q$ (see Note 2 before Example 11.6). By Problem 10 above, we have $U_f(R) \leq U_f(P)$ and $U_g(R) \leq U_g(Q)$. So, $U_f(P) + U_g(Q) \geq U_f(R) + U_g(R)$. Since $U_f(R) + U_g(R) \in C$, we see that $U_f(P) + U_g(Q) \geq U_f(R) + U_g(R) \geq \inf C$. Therefore, $U_g(Q) \geq \inf C - U_f(P)$. Since $U_g(Q) \in B$ was arbitrary, $\inf C - U_f(P)$ is a lower bound of B. Since $\inf B$ is the **greatest** lower bound of B, it follows that $\inf B \geq \inf C - U_f(P)$. Therefore, $U_f(P) \geq \inf C - \inf B$. Since $U_f(P) \in A$ was arbitrary, we see that $\inf C - \inf B$ is a lower bound of A. Since $\inf A$ is the **greatest** lower bound of A, we have $\inf A \geq \inf C - \inf B$. So, $\inf A + \inf B \geq \inf C$.

Since $\inf C \geq \inf A + \inf B$ and $\inf A + \inf B \geq \inf C$, we have $\inf C = \inf A + \inf B$. $\square$

LEVEL 5

13. Let $f: [a, b] \to \mathbb{R}$ be a monotonic function. Prove that f is Riemann integrable on $[a, b]$.

Proof: First let's assume that f is increasing on $[a, b]$. Then for all $x \in [a, b]$, $f(a) \leq f(x) \leq f(b)$. Therefore, f is bounded on $[a, b]$. We will use Lemma 11.10 to show that f is Riemann integrable on $[a, b]$.

So, let $\epsilon > 0$. By the Archimedean Property of $\mathbb{R}$ (Theorem 6.16), there is $n \in \mathbb{N}$ such that $n > \frac{(b-a)(f(b)-f(a))}{\epsilon}$. Let P be the partition consisting of n equal subintervals of $[a,b]$. In other words, if we let $x_i = a + \frac{i(b-a)}{n}$, then $P = \{[x_{i-1}, x_i]\}_{i=1}^n$. Then we have the following:

$$L_f(P) = \sum_{i=1}^n f(x_{i-1})\Delta x_i = \frac{b-a}{n}\sum_{i=1}^n f(x_{i-1})$$

$$U_f(P) = \sum_{i=1}^n f(x_i)\Delta x_i = \frac{b-a}{n}\sum_{i=1}^n f(x_i)$$

Therefore,

$$U_f(P) - L_f(P) = \frac{b-a}{n}\sum_{i=1}^n f(x_i) - \frac{b-a}{n}\sum_{i=1}^n f(x_{i-1})$$

$$= \frac{b-a}{n}\left(\sum_{i=1}^n (f(x_i) - f(x_{i-1}))\right) = \frac{(b-a)(f(b)-f(a))}{n} < \epsilon.$$

By Lemma 11.10, f is Riemann intagrable on $[a,b]$.

Next, let's assume that f is decreasing on $[a,b]$. Then for all $x \in [a,b]$, $f(b) \leq f(x) \leq f(a)$. Therefore, once again, f is bounded on $[a,b]$. We will again use Lemma 11.10 to show that f is Riemann integrable on $[a,b]$.

So, let $\epsilon > 0$. By the Archimedean Property of $\mathbb{R}$ (Theorem 6.16), there is $n \in \mathbb{N}$ such that $n > \frac{(b-a)(f(a)-f(b))}{\epsilon}$. Let P be the partition consisting of n equal subintervals of $[a,b]$. In other words, if we let $x_i = a + \frac{i(b-a)}{n}$, then $P = \{[x_{i-1}, x_i]\}_{i=1}^n$. Then we have the following:

$$L_f(P) = \sum_{i=1}^n f(x_i)\Delta x_i = \frac{b-a}{n}\sum_{i=1}^n f(x_i)$$

$$U_f(P) = \sum_{i=1}^n f(x_{i-1})\Delta x_i = \frac{b-a}{n}\sum_{i=1}^n f(x_{i-1})$$

Therefore,

$$U_f(P) - L_f(P) = \frac{b-a}{n}\sum_{i=1}^n f(x_{i-1}) - \frac{b-a}{n}\sum_{i=1}^n f(x_i)$$

$$= \frac{b-a}{n}\left(\sum_{i=1}^n (f(x_{i-1}) - f(x_i))\right) = \frac{(b-a)(f(a)-f(b))}{n} < \epsilon.$$

By Lemma 11.10, f is Riemann intagrable on $[a,b]$. $\qquad\square$

14. If P is a partition of $[a, b]$, we define the **norm** of P, written $\|P\|$ to be the maximum length of a subinterval in P. Prove that a bounded function $f: [a, b] \to \mathbb{R}$ is Riemann integrable if and only if there is $L \in \mathbb{R}$ such that the following holds: For every $\epsilon > 0$ there is $\delta > 0$ such that whenever P is a partition of $[a, b]$ with $\|P\| < \delta$ and S is any Riemann sum over P, then $|S - L| < \epsilon$. You may use Problem 19 below.

Proof: Assume that f is Riemann integrable on $[a, b]$, let $L = \int_a^b f(x)\, dx$, and let $\epsilon > 0$. By Problem 19, there is $\delta > 0$ such that whenever P is a partition with $\|P\| < \delta$, we have both $L_f(P) > L - \epsilon$ and $U_f(P) < L + \epsilon$. So, if $\|P\| < \delta$ and S is any Riemann sum over P, then we have

$$L - \epsilon < L_f(P) \leq S \leq U_f(P) < L + \epsilon.$$

So, $|S - L| < \epsilon$, as desired.

Now, assume that there is $L \in \mathbb{R}$ such that the following holds: For every $\epsilon > 0$ there is $\delta > 0$ such that whenever P is a partition with $\|P\| < \delta$ and S is any Riemann sum over P, then $|S - L| < \epsilon$ (or equivalently, $L - \epsilon < S < L + \epsilon$).

Let $\epsilon > 0$ be given, let $\delta > 0$ satisfy the given condition with ϵ replaced by $\frac{\epsilon}{2}$, let $P = \{[x_{i-1}, x_i]\}_{i=1}^n$ be a partition with $\|P\| < \delta$, and let $m_i = \inf\{f(x) \mid x_{i-1} \leq x \leq x_i\}$ and $M_i = \sup\{f(x) \mid x_{i-1} \leq x \leq x_i\}$ for each $i = 1, 2, \dots, n$.

Choose tags $c_1, c_2, \dots, c_n$ with $c_i \in [x_{i-1}, x_i]$ such that $f(c_i) > M_i - \frac{\epsilon}{2(b-a)}$ (we can do this because $M_i - \frac{\epsilon}{2(b-a)}$ is **not** an upper bound of $\{f(x) \mid x_{i-1} \leq x \leq x_i\}$) and tags $d_1, d_2, \dots, d_n$ with $d_i \in [x_{i-1}, x_i]$ such that $f(d_i) < m_i + \frac{\epsilon}{2(b-a)}$ (we can do this because $m_i + \frac{\epsilon}{2(b-a)}$ is **not** a lower bound of $\{f(x) \mid x_{i-1} \leq x \leq x_i\}$). Let S_U and S_L be the corresponding Riemann sums. Then we have:

$$S_U = \sum_{i=1}^n f(c_i)\Delta x_i > \sum_{i=1}^n \left(M_i - \frac{\epsilon}{2(b-a)} \right) \Delta x_i = \sum_{i=1}^n M_i \Delta x_i - \frac{\epsilon}{2(b-a)} \sum_{i=1}^n \Delta x_i$$

$$= U_f(P) - \frac{\epsilon}{2(b-a)}(b-a) = U_f(P) - \frac{\epsilon}{2} \geq \overline{\int_a^b} f(x)\, dx - \frac{\epsilon}{2}$$

$$S_L = \sum_{i=1}^n f(d_i)\Delta x_i < \sum_{i=1}^n \left(m_i + \frac{\epsilon}{2(b-a)} \right) \Delta x_i = \sum_{i=1}^n m_i \Delta x_i + \frac{\epsilon}{2(b-a)} \sum_{i=1}^n \Delta x_i$$

$$= L_f(P) + \frac{\epsilon}{2(b-a)}(b-a) = L_f(P) + \frac{\epsilon}{2} \leq \underline{\int_a^b} f(x)\, dx + \frac{\epsilon}{2}$$

The given condition tells us that $S_U < L + \frac{\epsilon}{2}$ and $S_L > L - \frac{\epsilon}{2}$. So, we have

$$\overline{\int_a^b} f(x)\, dx - \frac{\epsilon}{2} < S_U < L + \frac{\epsilon}{2} \quad \text{and} \quad \underline{\int_a^b} f(x)\, dx + \frac{\epsilon}{2} > S_L > L - \frac{\epsilon}{2}.$$

So, we have

$$\overline{\int_a^b} f(x)\, dx < L + \epsilon \quad \text{and} \quad \underline{\int_a^b} f(x)\, dx > L - \epsilon.$$

Also, we have

$$\overline{\int_a^b} f(x)\, dx \geq L \quad \text{and} \quad \underline{\int_a^b} f(x)\, dx \leq L \text{ (Check this!)}.$$

Since $\epsilon > 0$ was arbitrary, by Problem 15 from Problem Set 6, we have

$$\overline{\int_a^b} f(x)\, dx \leq L \quad \text{and} \quad \underline{\int_a^b} f(x)\, dx \geq L.$$

Therefore,

$$L \leq \underline{\int_a^b} f(x)\, dx \leq \overline{\int_a^b} f(x)\, dx \leq L.$$

So, $\underline{\int_a^b} f(x)\, dx = \overline{\int_a^b} f(x)\, dx$, and therefore, f is Riemann integrable on $[a,b]$. $\square$

15. Let $f: [a,b] \to \mathbb{R}$ be a bounded function that is Riemann integrable on $[a,b]$. Let $c, d, e \in [a,b]$. Prove that

$$\int_c^e f(x)\, dx = \int_c^d f(x)\, dx + \int_d^e f(x)\, dx.$$

Proof: We will first prove several lemmas:

Lemma 1: Let $a < c < b$ and let $f: [a,b] \to \mathbb{R}$ be bounded on $[a,b]$. Then

$$\overline{\int_a^b} f(x)\, dx = \overline{\int_a^c} f(x)\, dx + \overline{\int_c^b} f(x)\, dx \quad \text{and} \quad \underline{\int_a^b} f(x)\, dx = \underline{\int_a^c} f(x)\, dx + \underline{\int_c^b} f(x)\, dx.$$

Proof of Lemma 1: Let $Q = \{[x_{i-1}, x_i]\}_{i=1}^k$ be an arbitrary partition of $[a,b]$ and let $R = \{[y_{i-1}, y_i]\}_{i=1}^j \cup \{[y_{i-1}, y_i]\}_{i=j+1}^t$ be a refinement of Q with $y_j = c$. By Problem 10 above,

$$L_f(Q) \leq L_f(R)$$

$$\leq \sup\{L_f(P) \mid P = \{[x_{i-1}, x_i]\}_{i=1}^n \text{ is a partition of } [a,b] \text{ with } c = x_i \text{ for some } i = 1, 2, \ldots, n\}.$$

Since Q was an arbitrary partition of $[a,b]$, we see that

$$\sup\{L_f(P) \mid P \text{ is a partition of } [a,b]\}$$

$$\leq \sup\{L_f(P) \mid P = \{[x_{i-1}, x_i]\}_{i=1}^n \text{ is a partition of } [a,b] \text{ with } c = x_i \text{ for some } i = 1, 2, \ldots, n\}.$$

Now, if $Q = \{[x_{i-1}, x_i]\}_{i=1}^{k}$ is a partition of $[a, b]$ such that $c = x_i$ for some $i = 1, 2, ..., k$, then Q is a partition of $[a, b]$. It follows that

$$L_f(Q) \leq \sup\{L_f(P) \mid P \text{ is a partition of } [a, b]\}.$$

Since Q was an arbitrary partition of $[a, b]$ such that $c = x_i$ for some $i = 1, 2, ..., n$, we have

$$\sup\{L_f(P) \mid P = \{[x_{i-1}, x_i]\}_{i=1}^{n} \text{ is a partition of } [a, b] \text{ with } c = x_i \text{ for some } i = 1, 2, ..., n\}$$
$$\leq \sup\{L_f(P) \mid P \text{ is a partition of } [a, b]\}.$$

It follows that

$$\sup\{L_f(P) \mid P = \{[x_{i-1}, x_i]\}_{i=1}^{n} \text{ is a partition of } [a, b] \text{ with } c = x_i \text{ for some } i = 1, 2, ..., n\}$$
$$= \sup\{L_f(P) \mid P \text{ is a partition of } [a, b]\}.$$

Therefore, we have

$$\underline{\int_a^c} f(x)\, dx + \underline{\int_c^b} f(x)\, dx$$

$$= \sup\{L_f(P) \mid P \text{ a partition of } [a, c]\} + \sup\{L_f(Q) \mid Q \text{ a partition of } [c, d]\}$$
$$= \sup\{L_f(P) + L_f(Q) \mid P \text{ a partition of } [a, c] \wedge Q \text{ a partition of } [c, d]\} \text{ (by Problem 7 above)}$$
$$= \sup\{L_f(P) \mid P = \{[x_{i-1}, x_i]\}_{i=1}^{n} \text{ is a partition of } [a, b] \text{ with } c = x_i \text{ for some } i = 1, 2, ..., n\}$$
$$= \sup\{L_f(P) \mid P \text{ is a partition of } [a, b]\} = \underline{\int_a^b} f(x)\, dx.$$

An analogous argument shows that

$$\overline{\int_a^c} f(x)\, dx + \overline{\int_c^b} f(x)\, dx = \overline{\int_a^b} f(x)\, dx.$$

$\square$

Lemma 2: Let $a < c < b$ and let $f: [a, b] \to \mathbb{R}$ be bounded on $[a, b]$. Then f is Riemann integrable on $[a, b]$ if and only if f is Riemann integrable on $[a, c]$ and f is Riemann integrable on $[c, b]$.

Proof of Lemma 2: First assume that f is Riemann integrable on $[a, b]$. By Lemma 1 above, we have

$$\int_a^b f(x)\, dx = \underline{\int_a^b} f(x)\, dx = \underline{\int_a^c} f(x)\, dx + \underline{\int_c^b} f(x)\, dx$$

$$\leq \overline{\int_a^c} f(x)\, dx + \overline{\int_c^b} f(x)\, dx = \overline{\int_a^b} f(x)\, dx = \int_a^b f(x)\, dx.$$

So,

$$\underline{\int_a^c} f(x)\, dx + \underline{\int_c^b} f(x)\, dx = \overline{\int_a^c} f(x)\, dx + \overline{\int_c^b} f(x)\, dx.$$

116

We know that $\underline{\int_a^c} f(x)\, dx \leq \overline{\int_a^c} f(x)\, dx$. If we have $\underline{\int_a^c} f(x)\, dx < \overline{\int_a^c} f(x)\, dx$, then we must have

$$\underline{\int_c^b} f(x)\, dx = \left(\overline{\int_a^c} f(x)\, dx - \underline{\int_a^c} f(x)\, dx \right) + \overline{\int_c^b} f(x)\, dx > \overline{\int_c^b} f(x)\, dx.$$

This is impossible, and so, we must have $\underline{\int_a^c} f(x)\, dx = \overline{\int_a^c} f(x)\, dx$. It then follows that $\underline{\int_c^b} f(x)\, dx = \overline{\int_c^b} f(x)\, dx$. Therefore, f is Riemann integrable on $[a,c]$ and $[c,b]$.

Conversely, if f is Riemann integrable on $[a,c]$ and f is Riemann integrable on $[c,b]$, then we have

$$\underline{\int_a^b} f(x)\, dx = \underline{\int_a^c} f(x)\, dx + \underline{\int_c^b} f(x)\, dx = \int_a^c f(x)\, dx + \int_c^b f(x)\, dx$$

$$= \overline{\int_a^c} f(x)\, dx + \overline{\int_c^b} f(x)\, dx = \overline{\int_a^b} f(x)\, dx.$$

Therefore, f is Riemann integrable on $[a,b]$, as desired. $\quad\square$

Lemma 3: Let $a < c < b$ and let $f\colon [a,b] \to \mathbb{R}$ be Riemann integrable on $[a,b]$, $[a,c]$, and $[c,b]$. Then

$$\int_a^b f(x)\, dx = \int_a^c f(x)\, dx + \int_c^b f(x)\, dx.$$

Proof of Lemma 3: By Lemma 1,

$$\int_a^b f(x)\, dx = \underline{\int_a^b} f(x)\, dx = \underline{\int_a^c} f(x)\, dx + \underline{\int_c^b} f(x)\, dx = \int_a^c f(x)\, dx + \int_c^b f(x)\, dx.$$

$\quad\square$

Lemma 4: Let $f\colon [a,b] \to \mathbb{R}$ be Riemann integrable on $[a,b]$ and let $[c,d] \subseteq [a,b]$. Then f is Riemann integrable on $[c,d]$.

Proof of Lemma 4: By Lemma 2, f is Riemann integrable on $[a,c]$ and $[c,b]$. By Lemma 2 again, f is Riemann integrable on $[c,d]$ and $[d,b]$. In particular, f is Riemann integrable on $[c,d]$. $\quad\square$

Proof of main result: Let $c, d, e \in [a,b]$. We will prove the result by cases.

Case 1 ($c < d < e$): By Lemma 4, f is Riemann integrable on each of $[c,e]$, $[c,d]$, and $[d,e]$. So, by Lemma 3, we have

$$\int_c^e f(x)\, dx = \int_c^d f(x)\, dx + \int_d^e f(x)\, dx.$$

Case 2 ($c < e < d$): By Lemma 4, f is Riemann integrable on each of $[c,d]$, $[c,e]$, and $[e,d]$. So, by Lemma 3, we have

$$\int_c^d f(x)\,dx = \int_c^e f(x)\,dx + \int_e^d f(x)\,dx.$$

It follows that

$$\int_c^e f(x)\,dx = \int_c^d f(x)\,dx - \int_e^d f(x)\,dx = \int_c^d f(x)\,dx + \int_d^e f(x)\,dx.$$

Case 3 ($d < c < e$): By Lemma 4, f is Riemann integrable on each of $[d, e]$, $[d, c]$, and $[c, e]$. So, by Lemma 3, we have

$$\int_d^e f(x)\,dx = \int_d^c f(x)\,dx + \int_c^e f(x)\,dx.$$

It follows that

$$\int_c^e f(x)\,dx = \int_d^e f(x)\,dx - \int_d^c f(x)\,dx = \int_c^d f(x)\,dx + \int_d^e f(x)\,dx.$$

Case 4 ($d < e < c$): By Lemma 4, f is Riemann integrable on each of $[d, c]$, $[d, e]$, and $[e, c]$. So, by Lemma 3, we have

$$\int_d^c f(x)\,dx = \int_d^e f(x)\,dx + \int_e^c f(x)\,dx.$$

It follows that

$$\int_c^e f(x)\,dx = -\int_e^c f(x)\,dx = \int_d^e f(x)\,dx - \int_d^c f(x)\,dx = \int_c^d f(x)\,dx + \int_d^e f(x)\,dx.$$

Case 5 ($e < c < d$): By Lemma 4, f is Riemann integrable on each of $[e, d]$, $[e, c]$, and $[c, d]$. So, by Lemma 3, we have

$$\int_e^d f(x)\,dx = \int_e^c f(x)\,dx + \int_c^d f(x)\,dx.$$

It follows that

$$\int_c^e f(x)\,dx = -\int_e^c f(x)\,dx = \int_c^d f(x)\,dx - \int_e^d f(x)\,dx = \int_c^d f(x)\,dx + \int_d^e f(x)\,dx.$$

Case 6 ($e < d < c$): By Lemma 4, f is Riemann integrable on each of $[e, c]$, $[e, d]$, and $[d, c]$. So, by Lemma 3, we have

$$\int_e^c f(x)\,dx = \int_e^d f(x)\,dx + \int_d^c f(x)\,dx.$$

It follows that

$$\int_c^e f(x)\,dx = -\int_e^c f(x)\,dx = -\int_e^d f(x)\,dx - \int_d^c f(x)\,dx = \int_c^d f(x)\,dx + \int_d^e f(x)\,dx.$$

$\square$

16. Let $f:[a,b] \to \mathbb{R}$ be a bounded function with finitely many discontinuities. Prove that f is Riemann integrable on $[a,b]$.

Proof: We first prove a lemma:

Lemma: Let $f:[a,b] \to \mathbb{R}$ be a bounded function and for each $n \in \mathbb{N}$ let $[a_n, b_n]$ be a closed interval such that $a < a_n < b_n < b$ and f is Riemann integrable on $[a_n, b_n]$. If $\lim_{n\to\infty} a_n = a$ and $\lim_{n\to\infty} b_n = b$, then f is Riemann integrable on $[a,b]$.

Proof of Lemma: Since f is bounded on $[a,b]$, there is $M \in \mathbb{R}^+$ such that $-M \le f(x) \le M$ for all $x \in [a,b]$. Since $b - a > b_n - a_n$ for all $n \in \mathbb{N}$, by problem 6 above, we have

$$-M(b-a) < -M(b_n - a_n) = \int_{a_n}^{b_n} -M\,dx \le \int_{a_n}^{b_n} f(x)\,dx \le \int_{a_n}^{b_n} M\,dx = M(b_n - a_n) < M(b-a).$$

So, $\left(\int_{a_n}^{b_n} f(x)\,dx\right)$ is a bounded sequence of real numbers. By the Bolzano-Weierstrass Theorem (Theorem 8.27), $\left(\int_{a_n}^{b_n} f(x)\,dx\right)$ has a convergent subsequence $\left(\int_{a_{n_k}}^{b_{n_k}} f(x)\,dx\right)$. Let L be the limit of this subsequence.

By Lemma 1 in the solution to Problem 15 above and the fact that for each $n \in \mathbb{N}$, f is Riemann integrable on $[a_n, b_n]$, for each $k \in \mathbb{N}$, we have the following:

$$\overline{\int_a^b} f(x)\,dx = \overline{\int_a^{a_{n_k}}} f(x)\,dx + \overline{\int_{a_{n_k}}^{b_{n_k}}} f(x)\,dx + \overline{\int_{b_{n_k}}^b} f(x)\,dx$$

$$\le M(a_{n_k} - a) + \int_{a_{n_k}}^{b_{n_k}} f(x)\,dx + M(b - b_{n_k})$$

$$\underline{\int_a^b} f(x)\,dx = \underline{\int_a^{a_{n_k}}} f(x)\,dx + \underline{\int_{a_{n_k}}^{b_{n_k}}} f(x)\,dx + \underline{\int_{b_{n_k}}^b} f(x)\,dx$$

$$\ge -M(a_{n_k} - a) + \int_{a_{n_k}}^{b_{n_k}} f(x)\,dx - M(b - b_{n_k}).$$

Taking the limit as k goes to ∞, we get the following:

$$\overline{\int_a^b} f(x)\,dx \le M(a - a) + L + M(b - b) = M \cdot 0 + L + M \cdot 0 = L$$

$$\underline{\int_a^b} f(x)\,dx \ge -M(a - a) + L - M(b - b) = -M \cdot 0 + L - M \cdot 0 = L$$

So, we have

$$L \leq \int_{\underline{a}}^{b} f(x)\, dx \leq \overline{\int_{a}^{b}} f(x)\, dx \leq L.$$

Therefore, $\overline{\int_a^b} f(x)\, dx = \int_{\underline{a}}^{b} f(x)\, dx = L$. So, f is Riemann integrable on $[a,b]$. $\square$

Proof of main result: Let $f: [a,b] \to \mathbb{R}$ be bounded on $[a,b]$ with finitely many discontinuities. Let $P = \{[x_{i-1}, x_i]\}_{i=1}^{n}$ be a partition of $[a,b]$ such that f is continuous on (x_{i-1}, x_i) for each $i = 1, 2, \ldots, n$. If $x_{i-1} < y_{i-1} < y_i < x_i$, then f is continuous on $[y_{i-1}, y_i]$ and therefore, f is Riemann integrable on $[y_{i-1}, y_i]$. By the Lemma above, f is Riemann integrable on $[x_{i-1}, x_i]$. By Lemma 2 from Problem 15 above and a straightforward induction, f is Riemann integrable on $[a,b]$. $\square$

Problem Set 12

Full solutions to these problems are available for free download here:
www.SATPrepGet800.com/RABQXZ

LEVEL 1

1. Let $a \in (0, 1) \cup (1, \infty)$. Prove that for all $x \in (0, \infty)$, $\log_a x = \frac{\ln x}{\ln a}$.

Proof: Let $y = \log_a x$. Then $x = a^y$. So, $\ln x = \ln a^y = y \ln a$. Therefore, we have

$$\log_a x = y = \frac{\ln x}{\ln a}.$$

□

2. Let $a, b \in (0, 1) \cup (1, \infty)$. Prove that for all $x \in (0, \infty)$, $\log_a x = \frac{\log_b x}{\log_b a}$.

Proof: By Problem 1 above, we have $\log_b x = \frac{\ln x}{\ln b}$, or equivalently, $\ln x = (\ln b)(\log_b x)$.

Therefore, by Problem 1 again and the previous paragraph, we have

$$\log_a x = \frac{\ln x}{\ln a} = \frac{(\ln b)(\log_b x)}{\ln a} = \frac{\log_b x}{\frac{\ln a}{\ln b}} = \frac{\log_b x}{\log_b a}.$$

The first and last equalities follows from Problem 1 above.

□

3. Let $a \in (0, 1) \cup (1, \infty)$ and let $b, c \in (0, \infty)$. Prove that $\log_a bc = \log_a b + \log_a c$.

Proof: By Problem 1 above and Theorem 12.3, we have

$$\log_a bc = \frac{1}{\ln a} \ln bc = \frac{1}{\ln a}(\ln b + \ln c) = \frac{\ln b}{\ln a} + \frac{\ln c}{\ln a} = \log_a b + \log_a c.$$

□

4. Let $a \in (0, 1) \cup (1, \infty)$, let $b \in (0, \infty)$, and let $r \in \mathbb{R}$. Prove that $\log_a(b^r) = r \log_a b$.

Proof: By Problem 1 above and Theorem 12.17, we have

$$\log_a(b^r) = \frac{\ln(b^r)}{\ln a} = \frac{r \ln b}{\ln a} = r \cdot \frac{\ln b}{\ln a} = r \log_a b.$$

□

LEVEL 2

5. Use Problem 11 from Problem Set 11 to prove that $\frac{1}{3} \leq \ln 1.5 \leq \frac{1}{2}$.

121

Proof: We have

$$\ln 1.5 = \int_1^{1.5} \frac{1}{t}\, dt.$$

If $t \in [1, 1.5]$, then $\frac{1}{1.5} \leq \frac{1}{t} \leq 1$, or equivalently, $\frac{2}{3} \leq \frac{1}{t} \leq 1$ So, by Problem 11 from Problem Set 11 together with the fact that $f(x) = \frac{1}{x}$ is integrable on $[1, 1.5]$, we have $\frac{2}{3}(0.5) \leq \ln 1.5 \leq 1(0.5)$, or equivalently, $\frac{1}{3} \leq \ln 1.5 \leq \frac{1}{2}$. $\quad\square$

6. Use Riemann sums to prove that $\frac{25}{66} < \ln 1.5 < \frac{47}{110}$.

Proof: We have

$$\ln 1.5 = \int_1^{1.5} \frac{1}{t}\, dt.$$

Let $f(x) = \frac{1}{x}$ and let $P = \{[1, 1.125], [1.125, 1.25], [1.25, 1.375], [1.375, 1.5]\}$. Then we have

$$L_f(P) = \frac{1}{1.125} \cdot \frac{1}{8} + \frac{1}{1.25} \cdot \frac{1}{8} + \frac{1}{1.375} \cdot \frac{1}{8} + \frac{1}{1.5} \cdot \frac{1}{8} = \frac{763}{1980}$$

$$U_f(P) = \frac{1}{1} \cdot \frac{1}{8} + \frac{1}{1.125} \cdot \frac{1}{8} + \frac{1}{1.25} \cdot \frac{1}{8} + \frac{1}{1.375} \cdot \frac{1}{8} = \frac{1691}{3960}.$$

So, we see that $\frac{763}{1980} \leq \ln 2 \leq \frac{1691}{3960}$. Since $\frac{25}{66} = \frac{750}{1980} < \frac{763}{1980}$ and $\frac{1691}{3960} < \frac{1692}{3960} = \frac{47}{110}$, we see that $\frac{25}{66} < \ln 1.5 < \frac{47}{110}$. $\quad\square$

7. Use a geometric argument to prove that $\ln 1.5 < \frac{5}{12}$.

Proof: The area under the curve $f(x) = \frac{1}{x}$ between $x = 1$ and $x = 1.5$ is less than the area of the trapezoid formed with bases of length $f(1) = 1$ and $f(1.5) = \frac{1}{1.5} = \frac{2}{3}$ and height $1.5 - 1 = \frac{1}{2}$. This trapezoid has area $A = \frac{1}{2}(\text{base1} + \text{base2}) \cdot \text{height} = \frac{1}{2}\left(1 + \frac{2}{3}\right)\left(\frac{1}{2}\right) = \frac{5}{12}$. $\quad\square$

LEVEL 3

8. Let $x \in \mathbb{R}$. Prove that $e^x = \lim_{h \to 0}(1 + hx)^{\frac{1}{h}}$.

Proof: If $x = 0$, then $\lim_{h \to 0}(1 + hx)^{\frac{1}{h}} = \lim_{h \to 0}(1 + h \cdot 0)^{\frac{1}{h}} = \lim_{h \to 0} 1^{\frac{1}{h}} = \lim_{h \to 0} 1 = 1 = e^0$.

Now, assume that $x \neq 0$ and let $F(x) = \ln|x|$. Then $F'(x) = \frac{1}{x}$.

Using the definition of the derivative, the Note following Theorem 12.4, and Theorem 12.17, for $t \neq 0$, we have

$$\frac{1}{t} = F'(t) = \lim_{h \to 0} \frac{F(t+h) - F(t)}{h} = \lim_{h \to 0} \frac{\ln|t+h| - \ln|t|}{h} = \lim_{h \to 0} \frac{1}{h} \ln \frac{|t+h|}{|t|}$$

$$= \lim_{h \to 0} \frac{1}{h} \ln \left| \frac{t+h}{t} \right| = \lim_{h \to 0} \ln \left| \frac{t+h}{t} \right|^{\frac{1}{h}} = \lim_{h \to 0} \ln \left| 1 + \frac{h}{t} \right|^{\frac{1}{h}} = \lim_{h \to 0} \ln \left(1 + \frac{h}{t} \right)^{\frac{1}{h}}$$

We can drop the absolute value at the end because $1 + \frac{h}{t} > 0$ for h sufficiently close to 0. Specifically, we have $1 + \frac{h}{t} > 0$ whenever $\frac{h}{t} > -1$. If $t > 0$, this inequality is true for all $h > -t$ and if $t < 0$, this inequality is true for all $h < -t$.

Substituting $\frac{1}{x}$ for t gives us $x = \lim_{h \to 0} \ln(1 + hx)^{\frac{1}{h}}$.

Therefore, since e^x is continuous, $e^x = e^{\lim_{h \to 0} \ln(1+hx)^{\frac{1}{h}}} = \lim_{h \to 0} e^{\ln(1+hx)^{\frac{1}{h}}} = \lim_{h \to 0} (1 + hx)^{\frac{1}{h}}$. $\qquad \square$

9. Let $x \in \mathbb{R}$. Prove that $e^x = \lim_{h \to \infty} \left(1 + \frac{x}{h} \right)^h$.

Proof: By Problem 8 above, we have $e^x = \lim_{k \to 0} (1 + kx)^{\frac{1}{k}}$. It follows that $\lim_{k \to 0^+} (1 + kx)^{\frac{1}{k}} = e^x$.

If we let $h = \frac{1}{k}$, then $k = \frac{1}{h}$ and $h \to \infty$ if and only if $k \to 0^+$, and so, we have

$$\lim_{h \to \infty} \left(1 + \frac{x}{h} \right)^h = \lim_{k \to 0^+} (1 + kx)^{\frac{1}{k}} = e^x.$$

$\qquad \square$

10. Prove that for each $n \in \mathbb{Z}^+$ with $n \geq 2$,

$$\frac{1}{2} + \frac{1}{3} + \cdots + \frac{1}{n} < \ln n < 1 + \frac{1}{2} + \frac{1}{3} + \cdots + \frac{1}{n-1}.$$

Proof: Let $f(x) = \frac{1}{x}$, let $n \geq 2$ and let P be the partition of $[1, n]$ consisting of $n - 1$ equal subintervals of $[1, n]$. That is $P = \{[1, 2], [2, 3], \ldots, [n-1, n]\} = \{[i, i+1]\}_{i=1}^{n-1}$. Then, we have the following:

$$L_f(P) = \frac{1}{2} \cdot 1 + \frac{1}{3} \cdot 1 + \cdots + \frac{1}{n} \cdot 1 = \frac{1}{2} + \frac{1}{3} + \cdots + \frac{1}{n}$$

$$U_f(P) = 1 \cdot 1 + \frac{1}{2} \cdot 1 + \cdots + \frac{1}{n-1} \cdot 1 = 1 + \frac{1}{2} + \frac{1}{3} + \cdots + \frac{1}{n-1}$$

Since $\ln n = \int_1^n \frac{1}{t} \, dt$, we have $L_f(P) \leq \ln n \leq U_f(P)$. However, this isn't quite good enough. We need the inequalities to be strict.

Now, consider the partition Q of $[1, 2]$ consisting of two equal subintervals of $[1, 2]$. That is $Q = \{[1, 1.5], [1.5, 2]\}$, then we have seen in part 2 of Example 12.1 that $L_f(Q) = \frac{7}{12}$ and $U_f(Q) = \frac{5}{6}$. Let $R = Q \cup \{[2, 3], [3, 4] \ldots, [n-1, n]\}$. Then we have

123

$$L_f(R) = \frac{7}{12} + \frac{1}{3} + \cdots + \frac{1}{n} \quad \text{and} \quad U_f(R) = \frac{5}{6} + \frac{1}{2} + \cdots + \frac{1}{n-1}.$$

Notice how we can easily compute $L_f(R)$ and $U_f(R)$ by simply replacing the first terms of $L_f(P)$ and $U_f(P)$ by $\frac{7}{12}$ and $\frac{5}{6}$, respectively.

As before, we have $L_f(R) \leq \ln n \leq U_f(R)$.

Since $\frac{1}{2} = \frac{6}{12} < \frac{7}{12}$, we have $L_f(P) < L_f(R)$. Since $\frac{5}{6} < 1$, we have $U_f(R) < U_f(P)$.

Therefore, we have $L_f(P) < L_f(R) \leq \ln n \leq U_f(R) < U_f(P)$. $\qquad\square$

11. Prove that for all $x \in \mathbb{R}$, $1 + x \leq e^x$.

Proof: Let $f(x) = e^x - x - 1$. Then $f'(x) = e^x - 1$. Since e^x is strictly increasing on $\mathbb{R}$ it follows that if $x > 0$, then $e^x > e^0 = 1$, and consequently, for $x > 0$, $f'(x) > 1 - 1 = 0$. By part 1 of Corollary 10.26, f is strictly increasing on $[0, \infty)$. Since $f(0) = e^0 - 0 - 1 = 1 - 0 - 1 = 0$. It follows that for all $x > 0$, $f(x) > 0$, and therefore, $e^x > 1 + x$.

Once again, since e^x is strictly increasing on $\mathbb{R}$ it follows that if $x < 0$, then $e^x < e^0 = 1$, and consequently, for $x < 0$, $f'(x) < 1 - 1 = 0$. By part 2 of Corollary 10.26, f is strictly decreasing on $(-\infty, 0)$. Since $f(0) = 0$. It follows that for all $x < 0$, $f(x) > 0$, and therefore, $e^x > 1 + x$.

Finally, if $x = 0$, we have $1 + 0 = 1 = e^0$. $\qquad\square$

12. Prove that the following inequality holds:

$$\frac{4}{3} \leq \int_0^1 e^{x^2} \, dx \leq e.$$

Proof: By Problem 11 above, for all $x \in \mathbb{R}$, $1 + x^2 \leq e^{x^2}$. So, by Problem 6 in Problem Set 11,

$$\int_0^1 e^{x^2} \, dx \geq \int_0^1 (1 + x^2) \, dx = \left[x + \frac{x^3}{3} \right]_0^1 = 1 + \frac{1}{3} = \frac{4}{3}.$$

Now, since x^2 is increasing on $[0, 1]$ (because $\frac{d}{dx}[x^2] = 2x \geq 0$ for all $x \in [0, 1]$) and e^x is increasing on $[0, 1]$, we have e^{x^2} increasing on $[0, 1]$ (see the Note below for a detailed explanation). It follows that for all $x \in [0, 1]$, $e^{x^2} \leq e^{1^2} = e^1 = e$. So, by Problem 6 in Problem Set 11,

$$\int_0^1 e^{x^2} \, dx \leq \int_0^1 e \, dx = e \int_0^1 dx = e(1 - 0) = e. \qquad\square$$

Note: Let $f : [a, b] \to \mathbb{R}$ and $g : [c, d] \to \mathbb{R}$ be increasing functions with $f([a, b]) \subseteq [c, d]$. Let's verify carefully that $g \circ f : [a, b] \to \mathbb{R}$ is increasing. To this end, let $x, y \in [a, b]$ with $x < y$. Since f is increasing on $[a, b]$, $f(x) < f(y)$. Since $f([a, b]) \subseteq [c, d]$, we have $f(x), f(y) \in [c, d]$ with $f(x) < f(y)$. Since g is increasing on $[c, d]$, $(g \circ f)(x) = g(f(x)) < g(f(y)) = (g \circ f)(y)$, as desired.

13. Prove that for all $n \in \mathbb{Z}^+$ and all $x \in [0, \infty)$,

$$e^x \geq \sum_{k=0}^{n} \frac{x^k}{k!}$$

(where $0! = 1$ and $k! = 1 \cdot 2 \cdots k$ for all natural numbers $k \geq 1$).

Proof: For each $n \in \mathbb{Z}^+$, let

$$u_n(x) = \sum_{k=0}^{n} \frac{x^k}{k!}$$

We first prove by induction on n that for all $n \in \mathbb{N}$, $u'_{n+1} = u_n$.

Base case $(t = 0)$: $u'_{t+1}(x) = u'_1(x) = \frac{d}{dx}\left[\frac{x^0}{0!} + \frac{x^1}{1!}\right] = \frac{d}{dx}[1 + x] = 1 = \frac{x^0}{0!} = u_0(x) = u_t(x)$.

Inductive step: Assume that $t \geq 0$ and $u'_{t+1}(x) = u_t(x)$. Then we have

$$u'_{t+2}(x) = \frac{d}{dx}\left[u_{t+1}(x) + \frac{x^{t+2}}{(t+2)!}\right] = \frac{d}{dx}[u_{t+1}(x)] + \frac{d}{dx}\left[\frac{x^{t+2}}{(t+2)!}\right] = u_t(x) + \frac{x^{t+1}}{(t+1)!} = u_{t+1}(x).$$

If $x = 0$, we have

$$\sum_{k=0}^{t} \frac{x^k}{k!} = 1 + x + \frac{x^2}{2!} + \cdots + \frac{x^t}{t!} = 1 + 0 + \frac{0^2}{2!} + \cdots + \frac{0^t}{t!} = 1 = e^0.$$

We now prove by induction on $n \in \mathbb{Z}^+$ that for all $x \in (0, \infty)$,

$$e^x \geq \sum_{k=0}^{n} \frac{x^k}{k!}$$

For $t = 1$, we have $e^x \geq 1 + x = \frac{x^0}{0!} + \frac{x^1}{1!}$ by Problem 11 above. For the inductive step, assume $t \geq 1$ and that we have

$$e^x \geq \sum_{k=0}^{t} \frac{x^k}{k!} = u_t(x).$$

Let $f(x) = e^x - u_{t+1}(x)$. Then for all $x \in \mathbb{R}$, we have $f'(x) = e^x - u'_{t+1}(x) = e^x - u_t(x) \geq 0$. By parts (i) of Problem 10 in Problem Set 10, f is increasing on $\mathbb{R}$. Since $f(0) = e^0 - 1 = 1 - 1 = 0$. It follows that for all $x > 0$, $f(x) \geq 0$, and therefore, $e^x \geq u_{t+1}(x)$. □

14. For each $k = 1, 2, \ldots, n$, let $c_k \in (0, \infty)$, prove the following inequality:

$$(c_1 c_2 \cdots c_n)^{\frac{1}{n}} \leq \frac{c_1 + c_2 + \cdots + c_n}{n}$$

Proof: Let $M = \frac{c_1 + c_2 + \cdots + c_n}{n}$ and for each $k = 1, 2, \ldots, n$, let $x_k = \frac{c_k}{M} - 1$. By Problem 11 above, for each $k = 1, 2, \ldots, n$, we have

$$\frac{c_k}{M} = 1 + x_k \leq e^{x_k}.$$

Since the left- and right-hand sides of each of these n inequalities is positive, we can multiply to get

$$\frac{c_1}{M} \cdot \frac{c_2}{M} \cdots \frac{c_n}{M} \leq e^{x_1} e^{x_2} \cdots e^{x_n} = e^{x_1 + x_2 + \cdots + x_n} = e^{\frac{c_1 + c_2 + \cdots + c_n}{M} - n} = e^{n-n} = e^0 = 1.$$

Therefore, we have

$$c_1 c_2 \cdots c_n \leq M^n = \left(\frac{c_1 + c_2 + \cdots + c_n}{n} \right)^n.$$

Thus,

$$(c_1 c_2 \cdots c_n)^{\frac{1}{n}} \leq \frac{c_1 + c_2 + \cdots + c_n}{n}.$$

$\square$

15. Prove that for all $n \in \mathbb{Z}^+$ and all $x \in [0, \infty)$,

$$\ln(x + 1) = \sum_{k=1}^{n} \frac{(-1)^{k+1} x^k}{k} + \int_0^x \frac{(-1)^n t^n}{t + 1} \, dt.$$

Proof: If $n \in \mathbb{Z}^+$ and $x = 0$, then $\ln(0 + 1) = \ln 1 = 0$ and

$$\sum_{k=1}^{n} \frac{(-1)^{k+1} 0^k}{k} + \int_0^0 \frac{(-1)^n t^n}{t + 1} \, dt = 0 + 0 = 0.$$

Now, let $x \in (0, \infty)$. We prove by induction on $n \in \mathbb{Z}^+$ that

$$\frac{1 - (-1)^n x^n}{x + 1} = 1 - x + x^2 - x^3 + \cdots + (-1)^{n-1} x^{n-1}.$$

Base case ($k = 1$): We have

$$\frac{1 - (-1)^1 x^1}{x + 1} = \frac{1 - (-x)}{x + 1} = \frac{1 + x}{x + 1} = 1.$$

Inductive step: Assume that $k \in \mathbb{Z}^+$ and

$$\frac{1 - (-1)^k x^k}{x + 1} = 1 - x + x^2 - x^3 + \cdots + (-1)^{k-1} x^{k-1}$$

Then we have

$$1 - x + x^2 - x^3 + \cdots + (-1)^{k-1}x^{k-1} + (-1)^k x^k = \frac{1 - (-1)^k x^k}{x + 1} + (-1)^k x^k$$

$$= \frac{1 - (-1)^k x^k}{x + 1} + \frac{(-1)^k x^k (x + 1)}{x + 1} = \frac{1 - (-1)^k x^k + (-1)^k x^{k+1} + (-1)^k x^k}{x + 1}$$

$$= \frac{1 + (-1)^k x^{k+1}}{x + 1} = \frac{1 - (-1)^{k+1} x^{k+1}}{x + 1}.$$

So, by the Principle of Mathematical Induction, for all $n \in \mathbb{Z}^+$,

$$\frac{1}{x + 1} - \frac{(-1)^n x^n}{x + 1} = \frac{1 - (-1)^n x^n}{x + 1} = 1 - x + x^2 - x^3 + \cdots + (-1)^{n-1} x^{n-1},$$

or equivalently,

$$\frac{1}{x + 1} = 1 - x + x^2 - x^3 + \cdots + (-1)^{n-1} x^{n-1} + \frac{(-1)^n x^n}{x + 1}.$$

By the second form of the Fundamental Theorem of Calculus (Theorem 11.14), we have

$$\int_0^x \frac{1}{t + 1} \, dt = \int_0^x \left(1 - t + t^2 - t^3 + \cdots + (-1)^{n-1} t^{n-1} + \frac{(-1)^n t^n}{t + 1} \right) dt$$

$$= \left[t - \frac{t^2}{2} + \frac{t^3}{3} - \frac{t^4}{4} + \cdots + \frac{(-1)^{n-1} t^n}{n} \right]_0^x + \int_0^x \frac{(-1)^n t^n}{t + 1} \, dt$$

$$= x - \frac{x^2}{2} + \frac{x^3}{3} - \frac{x^4}{4} + \cdots + \frac{(-1)^{n+1} x^n}{n} + \int_0^x \frac{(-1)^n t^n}{t + 1} \, dt$$

$$= \sum_{k=1}^n \frac{(-1)^{k+1} x^k}{k} + \int_0^x \frac{(-1)^n t^n}{t + 1} \, dt.$$

Now, by definition, we have

$$\ln x = \int_1^x \frac{1}{t} \, dt.$$

We make the change of variables $u = t - 1$. Then $du = dt$, $u = 0$ when $t = 1$ and $u = x - 1$ when $t = x$. So, we have

$$\ln x = \int_0^{x-1} \frac{1}{u + 1} \, du.$$

Therefore, we have

$$\ln(x + 1) = \int_0^x \frac{1}{u + 1} \, du = \int_0^x \frac{1}{t + 1} \, dt = \sum_{k=1}^n \frac{(-1)^{k+1} x^k}{k} + \int_0^x \frac{(-1)^n t^n}{t + 1} \, dt.$$

$\square$

Problem Set 13

LEVEL 1

1. Prove that $\int_0^\infty \cos x \, dx$ diverges, but not to ∞ or $-\infty$.

Proof: If $c \geq 0$, then we have

$$\int_0^c \cos x \, dx = \sin c - \sin 0 = \sin c - 0 = \sin c.$$

Therefore,

$$\int_0^\infty \cos x \, dx = \lim_{c \to \infty} \int_0^c \cos x \, dx = \lim_{c \to \infty} \sin c.$$

Since $-1 \leq \sin c \leq 1$ for all $c \in \mathbb{R}$, $\lim_{c \to \infty} \sin c \neq \pm\infty$.

For every $n \in \mathbb{Z}$, we have $\sin n\pi = 0$ and $\sin \frac{(4n+1)\pi}{2} = 1$. It follows that $\lim_{c \to \infty} \sin c$ does not exist, and therefore, $\int_0^\infty \cos x \, dx$ diverges. $\square$

2. Let $a \in \mathbb{R} \cup \{-\infty\}$, let $b \in \mathbb{R} \cup \{\infty\}$, let $c \in (a, b)$, let $f: (a, b) \to \mathbb{R}$ be locally integrable on (a, b), and suppose that the integrals $\int_a^c f(x) \, dx$ and $\int_c^b f(x) \, dx$ are both finite real numbers. Prove that for all $d \in (a, b)$,

$$\int_a^c f(x) \, dx + \int_c^b f(x) \, dx = \int_a^d f(x) \, dx + \int_d^b f(x) \, dx.$$

Proof: Let F be an antiderivative of f on (a, b). By the second form of the Fundamental Theorem of Calculus (Theorem 11.14), we have

$$\int_a^c f(x) \, dx + \int_c^b f(x) \, dx = \lim_{k \to a^+} \int_k^c f(x) \, dx + \lim_{j \to b^-} \int_c^j f(x) \, dx$$

$$= \lim_{k \to a^+} \left(F(c) - F(k) \right) + \lim_{j \to b^-} \left(F(j) - F(c) \right) = F(c) - \lim_{k \to a^+} F(k) + \lim_{j \to b^-} F(j) - F(c)$$

$$= \lim_{j \to b^-} F(j) - \lim_{k \to a^+} F(k).$$

Since the improper integrals $\int_a^c f(x) \, dx$ and $\int_c^b f(x) \, dx$ are both finite real numbers, it follows that $\lim_{j \to b^-} F(j)$ and $\lim_{k \to a^+} F(k)$ both exist. Therefore,

$$\int_a^d f(x) \, dx + \int_d^b f(x) \, dx = \lim_{k \to a^+} \int_k^d f(x) \, dx + \lim_{j \to b^-} \int_d^j f(x) \, dx$$

$$= \lim_{k \to a^+} \left(F(d) - F(k) \right) + \lim_{j \to b^-} \left(F(j) - F(d) \right) = F(d) - \lim_{k \to a^+} F(k) + \lim_{j \to b^-} F(j) - F(d)$$

$$= \lim_{j \to b^-} F(j) - \lim_{k \to a^+} F(k) = \int_a^c f(x) \, dx + \int_c^b f(x) \, dx. \qquad \square$$

3. Let $c, d \in \mathbb{R}$, let I be an interval with endpoints a and b ($a < b$, a can be $-\infty$, and b can be ∞), and let $f, g: I \to \mathbb{R}$ be locally integrable on I such that the improper integrals of f and g over I converge. Prove that the improper integral of $cf + dg$ over I converges and

$$\int_a^b (cf + dg)(x)\, dx = c \int_a^b f(x)\, dx + d \int_a^b g(x)\, dx.$$

Proof: If $I = [a, b)$, then by Theorem 11.9 and Problem 9 from Problem Set 11, we have

$$\int_a^b (cf + dg)(x)\, dx = \lim_{k \to b^-} \int_a^k (cf + dg)(x)\, dx = \lim_{k \to b^-} \left[c \int_a^k f(x)\, dx + d \int_a^k g(x)\, dx \right]$$

$$= c \left[\lim_{k \to b^-} \int_a^k f(x)\, dx \right] + d \left[\lim_{k \to b^-} \int_a^k g(x)\, dx \right] = c \int_a^b f(x)\, dx + d \int_a^b g(x)\, dx.$$

If $I = (a, b]$, then again by Theorem 11.9 and Problem 9 from Problem Set 11, we have

$$\int_a^b (cf + dg)(x)\, dx = \lim_{k \to a^+} \int_k^b (cf + dg)(x)\, dx = \lim_{k \to a^+} \left[c \int_k^b f(x)\, dx + d \int_k^b g(x)\, dx \right]$$

$$= c \left[\lim_{k \to a^+} \int_k^b f(x)\, dx \right] + d \left[\lim_{k \to a^+} \int_k^b g(x)\, dx \right] = c \int_a^b f(x)\, dx + d \int_a^b g(x)\, dx.$$

If $I = (a, b)$, let $k \in (a, b)$. Then by the two previous results, we have

$$\int_a^b (cf + dg)(x)\, dx = \int_a^k (cf + dg)(x)\, dx + \int_k^b (cf + dg)(x)\, dx$$

$$= \left(c \int_a^k f(x)\, dx + d \int_a^k g(x)\, dx \right) + \left(c \int_k^b f(x)\, dx + d \int_k^b g(x)\, dx \right)$$

$$= \left(c \int_a^k f(x)\, dx + c \int_k^b f(x)\, dx \right) + \left(d \int_a^k g(x)\, dx + c \int_k^b f(x)\, dx \right)$$

$$= c \left(\int_a^k f(x)\, dx + \int_k^b f(x)\, dx \right) + d \left(\int_a^k g(x)\, dx + \int_k^b f(x)\, dx \right)$$

$$= c \int_a^b f(x)\, dx + d \int_a^b g(x)\, dx.$$

$\square$

4. For each $p \in \mathbb{R}$, evaluate the following improper integral:

$$\int_0^1 \frac{1}{x^p}\, dx$$

Solution: Let $p \in \mathbb{R}$. The function $f_p: (0, 1] \to \mathbb{R}$ defined by $f(x) = \frac{1}{x^p}$ is locally integrable on $(0, 1]$. First assume that $p \neq 1$. If $0 < c < 1$, then by the second form of the Fundamental Theorem of Calculus (Theorem 11.14), we have

$$\int_c^1 \frac{1}{x^p} \, dx = \int_c^1 x^{-p} \, dx = \left[\frac{x^{1-p}}{1-p} \right]_c^1 = \frac{1}{1-p} - \frac{c^{1-p}}{1-p} = \frac{1 - c^{1-p}}{1-p}.$$

It follows that the improper integral of $g(x) = \frac{1}{x^p}$ over $(0, 1]$ is

$$\int_0^1 \frac{1}{x^p} \, dx = \lim_{c \to 0^+} \int_0^1 \frac{1}{x^p} \, dx = \lim_{c \to 0^+} \left[\frac{1 - c^{1-p}}{1-p} \right] = \frac{1}{1-p} \lim_{c \to 0^+} [1 - c^{1-p}].$$

If $p < 1$, then $\lim_{c \to 0^+} [1 - c^{1-p}] = 1 - 0 = 1$, and so, the improper integral converges. In this case, we have

$$\int_0^1 \frac{1}{x^p} \, dx = \frac{1}{1-p} \lim_{c \to 0^+} [1 - c^{1-p}] = \frac{1}{1-p} \cdot 1 = \frac{1}{1-p}.$$

If $p > 1$, then $\lim_{c \to 0^+} [1 - c^{1-p}] = -\infty$. Since $\frac{1}{1-p} < 0$, we have

$$\int_0^1 \frac{1}{x^p} \, dx = \frac{1}{1-p} \lim_{c \to 0^+} [1 - c^{1-p}] = \infty \text{ (and so, the improper integral diverges)}.$$

If $p = 1$, then for $0 < c < 1$, we have

$$\int_c^1 \frac{1}{x^p} \, dx = \int_c^1 \frac{1}{x} \, dx = \ln 1 - \ln c = -\ln c.$$

It follows that the improper integral of $g(x) = \frac{1}{x}$ over $(0, 1]$ is

$$\int_0^1 \frac{1}{x} \, dx = \lim_{c \to 0^+} \int_0^1 \frac{1}{x} \, dx = \lim_{c \to 0^+} [-\ln c] = \infty \text{ (and so, the improper integral diverges)}.$$

$\square$

5. Use integration by parts to show that the following improper integral diverges:

$$\int_0^\infty \ln x \, dx$$

Solution: The function $f: (0, \infty) \to \mathbb{R}$ defined by $f(x) = \ln x$ is locally integrable on $(0, \infty)$. So, by definition, we have

$$\int_0^\infty \ln x \, dx = \int_0^1 \ln x \, dx + \int_1^\infty \ln x \, dx$$

To show that the integral on the left-hand side diverges, we need only show that one of the integrals on the right-hand side diverges. We will show that the second integral on the right-hand side diverges.

Let $u = \ln x$ and $dv = dx$. It then follows that $du = \frac{1}{x} dx$ and $v = x$. So, we have

$$\int_1^\infty \ln x \, dx = \lim_{c \to \infty} \int_1^c \ln x \, dx = \lim_{c \to \infty} \left([x \ln x]_1^c - \int_1^c dx \right) = \lim_{c \to \infty} \left(c \ln c - (c-1) \right)$$

$$= \lim_{c \to \infty} \left(c(\ln c - 1) + 1 \right) = \infty.$$

It follows that $\int_1^\infty \ln x \, dx$ diverges, and therefore, $\int_0^\infty \ln x \, dx$ diverges. $\quad\square$

Note: The first integral on the right-hand side converges. To see this, we use integration by parts again. Let $u = \ln x$ and $dv = dx$. It then follows that $du = \frac{1}{x} dx$ and $v = x$. So, we have

$$\int_0^1 \ln x \, dx = \lim_{c \to 0^+} \int_c^1 \ln x \, dx = \lim_{c \to 0^+} \left([x \ln x]_c^1 - \int_c^1 dx \right) = \lim_{c \to 0^+} \left(-c \ln c - (1-c) \right).$$

We can now use L'Hôpital's rule to get

$$\lim_{c \to 0^+} (-c \ln c) = \lim_{c \to 0^+} \frac{-\ln c}{\frac{1}{c}} = \lim_{c \to 0^+} \frac{-\frac{1}{c}}{-\frac{1}{c^2}} = \lim_{c \to 0^+} \left(\frac{1}{c} \cdot c^2 \right) = \lim_{c \to 0^+} c = 0.$$

It follows that

$$\int_0^1 \ln x \, dx = \lim_{c \to 0^+} \left(-c \ln c - (1-c) \right) = \lim_{c \to 0^+} (-c \ln c) - \lim_{c \to 0^+} (1-c) = 0 - 1 = -1.$$

6. Prove that for $p > 1$, $\int_1^\infty \frac{\sin x}{x^p} \, dx$ converges absolutely.

Proof: We have

$$\left| \frac{\sin x}{x^p} \right| = \frac{|\sin x|}{x^p} \le \frac{1}{x^p}.$$

By part 3 of Example 13.2, for $p > 1$, $\int_1^\infty \frac{1}{x^p} \, dx$ converges. Therefore, by the Comparison Test (Theorem 13.7), $\int_1^\infty \left| \frac{\sin x}{x^p} \right| \, dx$ converges. So, $\int_1^\infty \frac{\sin x}{x^p} \, dx$ converges absolutely. $\quad\square$

LEVEL 3

7. The function $\Gamma: (0, \infty) \to \mathbb{R}$ defined as follows is known as the **Gamma function**:

$$\Gamma(x) = \int_0^\infty t^{x-1} e^{-t} \, dt$$

Prove that for all $x \in (0, \infty)$, $\Gamma(x)$ converges.

Proof: Since $e^0 = 1$ and e^x is a strictly increasing function, for all $t > 0$, we have $e^t > 1$. Therefore, for all $t > 0$, $e^{-t} = \frac{1}{e^t} < 1$. So, for all $t > 0$, we have $t^{x-1}e^{-t} < t^{x-1}$. By the solution to Problem 4 above, for all $x > 0$, $\int_0^1 t^{x-1}\, dt = \int_0^1 \frac{1}{t^{1-x}}\, dt$ converges (if $x > 0$, then $1 - x < 1$). By the Comparison Test (Theorem 13.7), for all $x > 0$, $\int_0^1 t^{x-1}e^{-t}\, dt$ converges.

Now, for $x > 0$, we have $\lim_{t \to \infty} \frac{t^{x-1}e^{-t}}{\frac{1}{t^2}} = \lim_{t \to \infty} \frac{t^{x+1}}{e^t} = 0$ (Check this!). By part 3 of Example 13.2, $\int_1^\infty \frac{1}{t^2}\, dt$ converges. Therefore, by the Limit Comparison Test (Theorem 13.9), $\int_1^\infty t^{x-1}e^{-t}\, dt$ converges.

Since,

$$\Gamma(x) = \int_0^\infty t^{x-1}e^{-t}\, dt = \int_0^1 t^{x-1}e^{-t}\, dt + \int_1^\infty t^{x-1}e^{-t}\, dt,$$

it follows that $\Gamma(x)$ converges. $\qquad\square$

8. Let $a \in \mathbb{R}$, let $b \in \mathbb{R} \cup \{\infty\}$, and let $f: [a, b) \to \mathbb{R}$ be locally integrable on $[a, b)$. If $\int_a^b f(x)\, dx$ converges absolutely, then $\int_a^b f(x)\, dx$ converges.

Proof: Define $g: [a, b) \to \mathbb{R}$ by $g(x) = |f(x)| - f(x) = \begin{cases} 0 & \text{if } f(x) \geq 0. \\ -2f(x) & \text{if } f(x) < 0. \end{cases}$ Observe that for all $x \in [a, b)$, we have $0 \leq g(x) \leq 2|f(x)|$. Since $\int_a^b f(x)\, dx$ converges absolutely, $\int_a^b 2|f(x)|\, dx$ converges, and so, by the Comparison Test (Theorem 13.7), $\int_a^b g(x)\, dx$ converges. By Problem 3 above, $\int_a^b f(x)\, dx = \int_a^b (|f(x)| - g(x))\, dx = \int_a^b |f(x)|\, dx - \int_a^b g(x)\, dx$ converges. $\qquad\square$

9. Let $f: \left(-\frac{\pi}{2}, \frac{\pi}{2}\right) \to \mathbb{R}$ be defined by $f(x) = \tan x$.

    (i)   Prove that f is a continuous bijection from $\left(-\frac{\pi}{2}, \frac{\pi}{2}\right)$ onto $\mathbb{R}$.

    (ii) Let $g: \mathbb{R} \to \left(-\frac{\pi}{2}, \frac{\pi}{2}\right)$ be the inverse of f. We call the function g the arctangent function and we write $g(x) = \arctan x$. Prove that $g'(x) = \frac{1}{x^2+1}$ for all $x \in \mathbb{R}$.

    (iii) Determine if the following improper integral converges or diverges. If it converges, evaluate it.

    $$\int_{-\infty}^\infty \frac{1}{x^2 + 1}\, dx$$

Proofs:

    (i)   By Problem 22 in Problem Set 8, the functions $h, k: \mathbb{R} \to \mathbb{R}$ defined by $h(x) = \cos x$ and $k(x) = \sin x$ are continuous on $\mathbb{R}$. Also, $\cos x \neq 0$ for all $x \in \left(-\frac{\pi}{2}, \frac{\pi}{2}\right)$. By Problems 5 and 15 in Problem Set 8, $f(x) = \tan x = \frac{\sin x}{\cos x}$ is continuous on $\left(-\frac{\pi}{2}, \frac{\pi}{2}\right)$.

By the solution to Problem 3 in Problem Set 10, $f'(x) = \sec^2 x$, which is positive on $\left(-\frac{\pi}{2}, \frac{\pi}{2}\right)$. By part 1 of Corollary 10.26, f is strictly increasing on $\left(-\frac{\pi}{2}, \frac{\pi}{2}\right)$, and therefore, f is injective on $\left(-\frac{\pi}{2}, \frac{\pi}{2}\right)$. Indeed, if $x, y \in \left(-\frac{\pi}{2}, \frac{\pi}{2}\right)$ with $x \neq y$, then without loss of generality, we may assume that $x < y$. Since f is strictly increasing, $f(x) < f(y)$, and in particular, $f(x) \neq f(y)$.

We now show that $\lim\limits_{x \to \frac{\pi}{2}^-} \tan x = \infty$. To see this, first note that since $h(x) = \cos x$ is continuous on $\mathbb{R}$, $\lim\limits_{x \to \frac{\pi}{2}^-} \cos x = \cos\left(\frac{\pi}{2}\right) = 0$. Now, let $M > 0$. By the solution to part (i) of Problem 18 from Problem Set 8, we can choose $\delta > 0$ with $\delta < \frac{\pi}{4}$ (we can always choose δ smaller than necessary) so that $-\delta < x - \frac{\pi}{2} < 0$ implies $|\cos x - 0| < \frac{1}{M\sqrt{2}}$. Note that by insisting that $\delta < \frac{\pi}{4}$, we have $-\delta > -\frac{\pi}{4}$, and therefore, $\frac{\pi}{2} - \delta > \frac{\pi}{2} - \frac{\pi}{4} = \frac{\pi}{4}$. So, whenever $-\delta < x - \frac{\pi}{2} < 0$, or equivalently, $\frac{\pi}{2} - \delta < x < \frac{\pi}{2}$, we must have $x \in \left(\frac{\pi}{4}, \frac{\pi}{2}\right)$. Since for all $x \in \left(\frac{\pi}{4}, \frac{\pi}{2}\right)$, we have $\cos x > 0$, it follows that for all x with $\frac{\pi}{2} - \delta < x < \frac{\pi}{2}$, we have $\cos x < \frac{1}{M\sqrt{2}}$, or equivalently, $\frac{1}{\cos x} > M\sqrt{2}$. Since $\sin x > \frac{1}{\sqrt{2}}$ for all $x \in \left(\frac{\pi}{4}, \frac{\pi}{2}\right)$, we have

$$\tan x = \frac{\sin x}{\cos x} = \sin x \cdot \frac{1}{\cos x} > \frac{1}{\sqrt{2}} \cdot M\sqrt{2} = M.$$

By the solution to part (iv) of Problem 18 from Problem Set 8, $\lim\limits_{x \to \frac{\pi}{2}^-} \tan x = \infty$.

A similar argument shows that $\lim\limits_{x \to -\frac{\pi}{2}^+} \tan x = -\infty$.

Since $\lim\limits_{x \to \frac{\pi}{2}^-} \tan x = \infty$, $\lim\limits_{x \to -\frac{\pi}{2}^+} \tan x = -\infty$, and $f(x) = \tan x$ is continuous on $\left(-\frac{\pi}{2}, \frac{\pi}{2}\right)$, a straightforward application of the Intermediate Value Theorem (Corollary 9.9) shows that f is surjective. $\qquad\square$

(ii) Let $y = g(x) = \arctan x$. Then $\tan y = x$. Differentiating each side implicitly yields $(\sec^2 y)\frac{dy}{dx} = 1$. It follows that

$$g'(x) = \frac{dy}{dx} = \cos^2 y = \cos^2(\arctan x)$$

Now, $x = \tan y = \frac{\sin y}{\cos y}$. So, $x^2 + 1 = \frac{\sin^2 y}{\cos^2 y} + 1 = \frac{\sin^2 y}{\cos^2 y} + \frac{\cos^2 y}{\cos^2 y} = \frac{\sin^2 y + \cos^2 y}{\cos^2 y} = \frac{1}{\cos^2 y}$. It follows that $g'(x) = \cos^2(\arctan x) = \cos^2 y = \frac{1}{x^2+1}$. $\qquad\square$

(iii) We have the following:

$$\int_0^\infty \frac{1}{x^2+1}\,dx = \lim_{c\to\infty}\int_0^c \frac{1}{x^2+1}\,dx = \lim_{c\to\infty}[\arctan x]_0^c$$

$$= \lim_{c\to\infty}(\arctan c - \arctan 0) = \frac{\pi}{2} - 0 = \frac{\pi}{2}$$

$$\int_{-\infty}^0 \frac{1}{x^2+1}\,dx = \lim_{c\to-\infty}\int_c^0 \frac{1}{x^2+1}\,dx = \lim_{c\to-\infty}[\arctan x]_c^0$$

$$= \lim_{c\to-\infty}(\arctan 0 - c) = 0 - \left(-\frac{\pi}{2}\right) = \frac{\pi}{2}$$

Therefore, we have

$$\int_{-\infty}^\infty \frac{1}{x^2+1}\,dx = \int_{-\infty}^0 \frac{1}{x^2+1}\,dx + \int_0^\infty \frac{1}{x^2+1}\,dx = \frac{\pi}{2} + \frac{\pi}{2} = \pi.$$

$\square$

LEVEL 4

10. Let $a \in \mathbb{R}$, let $b \in \mathbb{R} \cup \{\infty\}$, let $f:[a,b) \to \mathbb{R}$ be a bounded locally integrable function on $[a,b)$ such that for all $x \in [a,b)$, $f(x) \geq 0$. Prove that

$$\int_a^b f(x)\,dx = \sup\{\textstyle\int_a^x f(t)\,dt \mid x \in [a,b)\}.$$

Proof: Define $F:[a,b) \to \mathbb{R}$ by $F(x) = \int_a^x f(t)\,dt$. Then for all $x \in [a,b)$, $F'(x) = f(x) \geq 0$. Therefore, by part (i) of Problem 10 from Problem Set 10, F is increasing on $[a,b)$.

There are four cases to consider.

Case 1 ($b = \infty$ and $\sup\{\int_a^x f(t)\,dt \mid x \in [a,\infty)\} = \infty$): Since $\sup\{\int_a^x f(t)\,dt \mid x \in [a,\infty)\} = \infty$, given $M \in \mathbb{R}$, there is $K \in [a,\infty)$ such that $F(K) = \int_a^K f(t)\,dt > M$. Since F is increasing on $[a,\infty)$, for all $x \in [a,\infty)$ with $x > K$, $F(x) \geq F(K) > M$. It follows from the solution to part (iv) of Problem 16 from Problem Set 8 that $\int_a^\infty f(x)\,dx = \lim_{x\to\infty}\int_a^x f(t)\,dt = \lim_{x\to\infty}F(x) = \infty$.

Case 2 ($b \in \mathbb{R}$ and $\sup\{\int_a^x f(t)\,dt \mid x \in [a,b)\} = \infty$): Since $\sup\{\int_a^x f(t)\,dt \mid x \in [a,b)\} = \infty$, given $M \in \mathbb{R}$, there is $K \in [a,b)$ such that $F(K) > M$. Since F is increasing on $[a,b)$, for all $x \in [a,b)$ with $x > K$, $F(x) \geq F(K) > M$. Let $\delta = b - K$. If $-\delta < x - b < 0$, then $b - \delta < x < b$, or equivalently, $K < x < b$, and so, $F(x) > M$. It follows from the solution to part (iv) of Problem 18 from Problem Set 8 that $\int_a^b f(x)\,dx = \lim_{x\to b^-}\int_a^x f(t)\,dt = \lim_{x\to b^-}F(x) = \infty$.

134

Case 3 ($b = \infty$ and $\sup\{\int_a^x f(t)\,dt \mid x \in [a, \infty)\} = L \in \mathbb{R}$): Let $\epsilon > 0$. Since $L - \epsilon < L$, it follows that $L - \epsilon$ is **not** an upper bound of $\{\int_a^x f(t)\,dt \mid x \in [a, \infty)\}$. Therefore, there is $K \in [a, \infty)$ such that $F(K) > L - \epsilon$. Since F is increasing on $[a, \infty)$, for all $x \in [a, \infty)$ with $x > K$, $F(x) \geq F(K) > L - \epsilon$. So, for all $x > K$, we have $L - F(x) < \epsilon$. Since L is an upper bound of the set $\{F(x) \mid x \in [a, b)\}$, for all $x \in [a, b)$, $L - F(x) \geq 0$. So, for all $x > K$, we have $|F(x) - L| = L - F(x) < \epsilon$. By the solution to part (ii) of Problem 16 from Problem Set 8, $\int_a^\infty f(x)\,dx = \lim_{x \to \infty} \int_a^x f(t)\,dt = \lim_{x \to \infty} F(x) = L$.

Case 4 ($b \in \mathbb{R}$ and $\sup\{\int_a^x f(t)\,dt \mid x \in [a, \infty)\} = L \in \mathbb{R}$): Let $\epsilon > 0$. Since $L - \epsilon < L$, it follows that $L - \epsilon$ is **not** an upper bound of $\{\int_a^x f(t)\,dt \mid x \in [a, b)\}$. Therefore, there is $K \in [a, b)$ such that $F(K) > L - \epsilon$. Since F is increasing on $[a, b)$, for all $x \in [a, b)$ with $x > K$, $F(x) \geq F(K) > L - \epsilon$. So, for all $x > K$, we have $L - F(x) < \epsilon$. Let $\delta = b - K$. If $-\delta < x - b < 0$, then $b - \delta < x < b$, or equivalently, $K < x < b$, and so, $L - F(x) < \epsilon$. Since L is an upper bound of the set $\{F(x) \mid x \in [a, b)\}$, for all $x \in [a, b)$, $L - F(x) \geq 0$. Therefore, we have $|F(x) - L| = L - F(x) < \epsilon$. So, we showed $\forall \epsilon > 0\, \exists \delta > 0\, (-\delta < x - b < 0 \to |F(x) - L| < \epsilon)$. By the solution to part (i) of Problem 18 from Problem Set 8, $\int_a^b f(x)\,dx = \lim_{x \to b^-} \int_a^x f(t)\,dt = \lim_{x \to b^-} F(x) = L$. $\qquad\square$

11. Prove that for $0 < p \leq 1$, $\int_1^\infty \frac{\sin x}{x^p}\,dx$ converges conditionally.

Proof: Let $0 < p \leq 1$. We first use integration by parts (Theorem 13.3) to show that the integral converges. Let $u(x) = x^{-p}$ and $dv = \sin x\,dx$. Then $du = -px^{-p-1} = -\frac{p}{x^{p+1}}$ and $v(x) = -\cos x\,dx$. So, we have

$$\int_1^\infty \frac{\sin x}{x^p}\,dx = \lim_{c \to \infty} \int_1^c \frac{\sin x}{x^p}\,dx = \lim_{c \to \infty}\left(\left[-\frac{\cos x}{x^p}\right]_1^c - \int_1^c \frac{p\cos x}{x^{p+1}}\,dx\right)$$

$$= \lim_{c \to \infty}\left(\cos 1 - \frac{\cos c}{c^p}\right) - p\int_1^\infty \frac{\cos x}{x^{p+1}}\,dx = \cos 1 - \lim_{c \to \infty}\frac{\cos c}{c^p} - p\int_1^\infty \frac{\cos x}{x^{p+1}}\,dx.$$

Since $-1 \leq \cos c \leq 1$ for all $c \geq 1$ (and in fact for all $c \in \mathbb{R}$), we have $-\frac{1}{c^p} \leq \frac{\cos c}{c^p} \leq \frac{1}{c^p}$ for all $c \geq 1$. We also have $\lim_{c \to \infty}\frac{1}{c^p} = 0$ and $\lim_{c \to \infty}\left[-\frac{1}{c^p}\right] = -\lim_{c \to \infty}\left[\frac{1}{c^p}\right] = 0$. Therefore, by an infinite version of the Squeeze Theorem we have $\lim_{c \to \infty}\frac{\cos c}{c^p} = 0$ (the dedicated reader may want to prove Problem 14 from Problem Set 8 in the case where $r = \infty$).

Also, for $x \geq 1$, we have

$$\left|\frac{\cos x}{x^{p+1}}\right| = \frac{|\cos x|}{x^{p+1}} \leq \frac{1}{x^{p+1}}.$$

Since $p > 0$, we have $p + 1 > 1$, and so, by part 3 of Example 13.2, $\int_1^\infty \frac{1}{x^{p+1}}\,dx$ converges. It follows that $\int_1^\infty \frac{\cos x}{x^{p+1}}\,dx$ converges absolutely. By Problem 8 above, $\int_1^\infty \frac{\cos x}{x^{p+1}}\,dx$ converges.

Therefore, $\int_1^\infty \frac{\sin x}{x^p}\,dx$ converges.

We now show that $\int_1^\infty \left|\frac{\sin x}{x^p}\right| dx$ diverges. We first consider the case $p = 1$.

Let $k \in \mathbb{N}$ with $k > 3$. We have

$$\int_1^{k\pi} \left|\frac{\sin x}{x}\right| dx = \int_1^{k\pi} \frac{|\sin x|}{x} dx > \int_\pi^{k\pi} \frac{|\sin x|}{x} dx = \sum_{i=1}^{k-1}\left(\int_{i\pi}^{(i+1)\pi} \frac{|\sin x|}{x} dx \right)$$

$$> \sum_{i=1}^{k-1}\left(\int_{i\pi}^{(i+1)\pi} \frac{|\sin x|}{(i+1)\pi} dx \right) = \sum_{i=1}^{k-1}\left(\frac{1}{(i+1)\pi} \int_{i\pi}^{(i+1)\pi} |\sin x|\, dx \right)$$

$$= \frac{1}{\pi}\sum_{i=1}^{k-1}\left(\frac{1}{(i+1)} \int_0^\pi \sin x\, dx \right) = \frac{1}{\pi}\sum_{i=1}^{k-1}\left(\frac{1}{(i+1)} [-\cos x]_0^\pi \right)$$

$$= \frac{1}{\pi}\sum_{i=1}^{k-1}\left(\frac{1}{(i+1)}(-\cos\pi + \cos 0) \right) = \frac{1}{\pi}\sum_{i=1}^{k-1}\left(\frac{1}{(i+1)} \cdot 2 \right) = \frac{2}{\pi}\sum_{i=1}^{k-1} \frac{1}{(i+1)}$$

$$= \frac{2}{\pi}\sum_{i=1}^{k-1} \int_{i+1}^{i+2} \frac{1}{(i+1)} dx > \frac{2}{\pi}\sum_{i=1}^{k-1} \int_{i+1}^{i+2} \frac{1}{x} dx = \frac{2}{\pi} \int_2^{k+1} \frac{1}{x} dx$$

$$= \frac{2}{\pi}(\ln(k+1) - \ln 2) = \frac{2}{\pi}\ln\frac{k+1}{2}.$$

Therefore,

$$\int_1^\infty \left|\frac{\sin x}{x^p}\right| dx = \lim_{k\to\infty} \int_1^{k\pi} \left|\frac{\sin x}{x^p}\right| dx > \lim_{k\to\infty}\left(\frac{2}{\pi}\ln\frac{k+1}{2} \right) = \infty.$$

So, $\int_1^\infty \left|\frac{\sin x}{x^p}\right| dx$ diverges.

Now, if $0 < p < 1$, then for $x \geq 1$, we have $0 \leq x^p < x$, and so, $0 \leq \frac{|\sin x|}{x} < \frac{|\sin x|}{x^p}$. Therefore, by the Comparison Test (Theorem 13.7), $\int_1^\infty \left|\frac{\sin x}{x^p}\right| dx$ diverges.

It follows that for $0 < p \leq 1$, $\int_1^\infty \frac{\sin x}{x^p} dx$ does **not** converge absolutely. So, $\int_1^\infty \frac{\sin x}{x^p} dx$ converges conditionally. □

12. Let $n \in \mathbb{Z}^+$ and let $f, g: [a, b] \to \mathbb{R}$ be continuous functions such that $f^{(k)}$ exists for each $k = 1, 2, \ldots, n$, where $f^{(k)}$ is the kth derivative of f. For each $k = 1, 2, \ldots, n$, let $g^{(-k)}$ be a kth antiderivative of g (in other words, $g^{(-(k+1))}$ is an antiderivative of $g^{(-k)}$). Prove that the following formula holds, where $f^{(0)} = f$:

$$\int_a^b f(x)g(x)\, dx = \left[\sum_{k=0}^{n-1} (-1)^k f^{(k)}(x) g^{(-(k+1))}(x) \right]_a^b + (-1)^n \int_a^b f^{(n)}(x) g^{(-n)}(x)\, dx$$

This formula is called **tabular integration by parts**.

Proof: We prove this by induction on $n \in \mathbb{Z}^+$.

Base Case $(t = 1)$: For $t = 1$, the formula is

$$\int_a^b f(x)g(x)\, dx = \left[(-1)^0 f^{(0)}(x) g^{(-1)}(x) \right]_a^b + (-1)^1 \int_a^b f^{(1)}(x) g^{(-1)}(x)\, dx$$

$$= f(b)g^{(-1)}(b) - f(a)g^{(-1)}(a) - \int_a^b f'(x)g^{(-1)}(x)\, dx$$

Setting $u(x) = f(x)$ and $v'(x) = g(x)$, we get

$$\int_a^b u(x)v'(x)\, dx = u(b)v(b) - u(a)v(a) - \int_a^b u'(x)v(x)\, dx$$

This is the integration by parts formula given in Theorem 13.3.

Inductive step: Let $t \in \mathbb{Z}^+$ with $t \geq 1$ and assume that the following is true:

$$\int_a^b f(x)g(x)\, dx = \left[\sum_{k=0}^{t-1} (-1)^k f^{(k)}(x) g^{(-(k+1))}(x) \right]_a^b + (-1)^t \int_a^b f^{(t)}(x) g^{(-t)}(x)\, dx$$

Let $u(x) = f^{(t)}(x)$ and $dv = g^{(-t)}(x)dx$. Then $du = f^{(t+1)}(x)$ and $v(x) = g^{(-(t+1))}(x)$. So, by the integration by parts formula given in Theorem 13.3, we have

$$\int_a^b f^{(t)}(x) g^{(-t)}(x)\, dx = \left[f^{(t)}(x) g^{(-(t+1))}(x) \right]_a^b - \int_a^b f^{(t+1)}(x) g^{(-(t+1))}(x)\, dx.$$

Therefore, we have

$$\int_a^b f(x)g(x)\ dx = \left[\sum_{k=0}^{t-1}(-1)^k f^{(k)}(x)g^{(-(k+1))}(x)\right]_a^b + (-1)^t \int_a^b f^{(t)}(x)g^{(-t)}(x)\ dx$$

$$= \left[\sum_{k=0}^{t-1}(-1)^k f^{(k)}(x)g^{(-(k+1))}(x)\right]_a^b + (-1)^t\left(\left[f^{(t)}(x)g^{(-(t+1))}(x)\right]_a^b - \int_a^b f^{(t+1)}(x)g^{(-(t+1))}(x)\ dx\right)$$

$$= \left[\sum_{k=0}^{t-1}(-1)^k f^{(k)}(x)g^{(-(k+1))}(x)\right]_a^b + (-1)^t\left[f^{(t)}(x)g^{(-(t+1))}(x)\right]_a^b + (-1)^t\left(-\int_a^b f^{(t+1)}(x)g^{(-(t+1))}(x)\ dx\right)$$

$$= \left[\sum_{k=0}^{t}(-1)^k f^{(k)}(x)g^{(-(k+1))}(x)\right]_a^b + (-1)^{t+1}\int_a^b f^{(t+1)}(x)g^{(-(t+1))}(x)\ dx$$

By the Principle of Mathematical Induction, the result holds. $\qquad\square$

13. Let $a \in \mathbb{R}$, let $b \in \mathbb{R}\cup\{\infty\}$, and let $f, g\colon [a,b) \to \mathbb{R}$ be functions such that f is continuous on $[a,b)$, g is differentiable on $[a,b)$, g' is absolutely integrable on $[a,b)$, and $\lim_{x\to b^-} g(x) = 0$. Suppose also that the function $F\colon [a,b) \to \mathbb{R}$ defined by $F(x) = \int_a^x f(t)\,dt$ is bounded on $[a,b)$. Prove that $\int_a^b f(x)g(x)\,dx$ converges. This result is known as **Dirichlet's Test** (Theorem 13.12).

Proof: Since g is differentiable on $[a,b)$, by Theorem 10.11, g is continuous on $[a,b)$. So, by Theorem 8.14 and Problem 5 from Problem Set 8, fg is continuous on $[a,b)$. By Theorem 11.7, fg is locally integrable on $[a,b)$. For $x \in [a,b)$, let $u(x) = g(x)$ and $v'(x) = f(x)$. Then $u'(x) = g'(x)$ and by the first form of the Fundamental Theorem of Calculus (Theorem 11.12), $v(x) = F(x)$. Using integration by parts (Theorem 13.3), for any $c \in [a,b)$, we have

$$\int_a^c f(x)g(x)\,dx = \int_a^c g(x)f(x)\,dx = \int_a^c u(x)v'(x)\,dx = u(c)v(c) - u(a)v(a) - \int_a^c v(x)u'(x)\,dx$$

$$= g(c)F(c) - g(a)F(a) - \int_a^c F(x)g'(x)\,dx = g(c)F(c) - 0 - \int_a^c F(x)g'(x)\,dx.$$

Therefore,

$$\int_a^b f(x)g(x)\,dx = \lim_{c\to b^-} g(c)F(c) - \int_a^b F(x)g'(x)\,dx.$$

Now, since $F(x)$ is bounded on $[a,b)$, there is $M \in \mathbb{R}^+$ such that $|F(x)| \le M$. It follows that for all $x \in [a,b)$, we have $|F(x)g'(x)| = |F(x)||g'(x)| \le M|g'(x)|$. Since g' is absolutely integrable on $[a,b)$, by the Comparison Test (Theorem 13.7), Fg' is absolutely integrable on $[a,b)$. So, by Problem 8 above, $\int_a^b F(x)g'(x)\,dx$ converges.

Since $|F(x)| \leq M$ for all $x \in [a, b)$, we have $|F(x)g(x)| \leq M|g(x)|$ for all $x \in [a, b)$, or equivalently $-M|g(x)| \leq F(x)g(x) \leq M|g(x)|$ for all $x \in [a, b)$. Since $\lim_{c \to b^-} g(c) = 0$, by the Squeeze Theorem (Problem 14 from Problem Set 8), $\lim_{c \to b^-} F(c)g(c) = 0$. So,

$$\int_a^b f(x)g(x)\,dx = 0 - \int_a^b F(x)g'(x)\,dx = -\int_a^b F(x)g'(x)\,dx.$$

Therefore, $\int_a^b f(x)g(x)\,dx$ converges. $\qquad \square$

Note: For the case $b = \infty$, you will need the infinite version of the Squeeze Theorem. The dedicated reader should prove this.

14. Let $a \in \mathbb{R}$, let $b \in \mathbb{R} \cup \{\infty\}$, and let $f, g : [a, b) \to \mathbb{R}$ be locally integrable on $[a, b)$. Assume that there is a subinterval $[c, b) \subseteq [a, b)$ such that for all $x \in [c, b)$, we have $f(x) \geq 0$ and $g(x) > 0$. Also assume that there is $L \in \mathbb{R} \cup (\infty)$ such that

$$\lim_{x \to b^-} \frac{f(x)}{g(x)} = L.$$

Prove each of the following:

$$\text{if } L \in \mathbb{R}^+, \text{then } \int_a^b f(x)\,dx \downarrow \text{ if and only if } \int_a^b g(x)\,dx \downarrow.$$

$$\text{if } L = 0 \text{ and } \int_a^b g(x)\,dx \downarrow, \text{then } \int_a^b f(x)\,dx \downarrow.$$

$$\text{if } L = \infty \text{ and } \int_a^b g(x)\,dx = \infty, \text{then } \int_a^b f(x)\,dx = \infty.$$

This result is known as the **Limit Comparison Test** (Theorem 13.9).

Proof: First assume that $L \in \mathbb{R}^+$. Since $\lim_{x \to b^-} \frac{f(x)}{g(x)} = L$, there is $d \in [c, b)$ such that

$$d < x < b \to \left| \frac{f(x)}{g(x)} - L \right| < \frac{L}{2},$$

or equivalently,

$$d < x < b \to 0 < \frac{L}{2} < \frac{f(x)}{g(x)} < \frac{3L}{2}.$$

Since $g(x) > 0$ for all $x \in [c, b)$, we have $d < x < b \to \frac{L}{2}g(x) < f(x) < \frac{3L}{2}g(x)$.

If $\int_a^b f(x)\,dx \downarrow$, then by the Comparison Test (Theorem 13.7) and the inequality $\frac{L}{2}g(x) < f(x)$, we have that $\int_a^b g(x)\,dx \downarrow$. Similarly, if $\int_a^b g(x)\,dx \downarrow$, then by the Comparison Test and the inequality $f(x) < \frac{3L}{2}g(x)$, we have that $\int_a^b f(x)\,dx \downarrow$.

Next, assume that $L = 0$. Since $\lim\limits_{x \to b^-} \frac{f(x)}{g(x)} = 0$, there is $d \in [c, b)$ such that

$$d < x < b \to 0 \leq \frac{f(x)}{g(x)} < 1.$$

Since $g(x) > 0$ for all $x \in [c, b)$, we have $d < x < b \to 0 \leq f(x) \leq g(x)$.

If $\int_a^b g(x)\, dx \downarrow$, then by the Comparison Test and the inequality $f(x) \leq g(x)$, we have that $\int_a^b f(x)\, dx \downarrow$.

Finally, assume that $L = \infty$. Since $\lim\limits_{x \to b^-} \frac{f(x)}{g(x)} = \infty$, there is $d \in [c, b)$ such that

$$d < x < b \to 0 < f(x) \text{ and } \frac{f(x)}{g(x)} \geq 1, \text{ or equivalently, } f(x) \geq g(x).$$

If $\int_a^b g(x)\, dx \uparrow$, then by the Comparison Test and the inequality $f(x) \geq g(x)$, we have that $\int_a^b f(x)\, dx \uparrow$. $\qquad\square$

Problem Set 14

LEVEL 1

1. Let (s_n) be a convergent sequence. Prove that $\lim_{n \to \infty} s_n$ is unique.

Proof: Suppose that $\lim_{n \to \infty} s_n = s$ and $\lim_{n \to \infty} s_n = t$. Let $\epsilon > 0$. Since $\lim_{n \to \infty} s_n = s$, we can find $K_1 \in \mathbb{N}$ such that $n > K_1 \to |s_n - s| < \frac{\epsilon}{2}$. Since $\lim_{n \to \infty} s_n = t$, we can find $K_2 \in \mathbb{N}$ such that $n > K_2 \to |s_n - t| < \frac{\epsilon}{2}$. . Let $K = \max\{K_1, K_2\}$. Suppose that $n > K$. Then $n > K_1$ and $n > K_2$. Therefore, we have

$$|s - t| = |(s_n - t) - (s_n - s)| \text{ (SACT)} \leq |s_n - t| + |s_n - s| \text{ (TI)} < \frac{\epsilon}{2} + \frac{\epsilon}{2} = \epsilon.$$

Since ϵ was an arbitrary positive real number, by Problem 15 from Problem Set 6, $|s - t| = 0$. So, $s - t = 0$, and therefore, $s = t$. $\quad\square$

2. Prove that a convergent sequence is bounded.

Proof: Let (s_n) be a convergent sequence and let $\lim_{n \to \infty} s_n = s$. Then for every $\epsilon > 0$, there is $K \in \mathbb{N}$ such that $n > K$ implies $|s_n - s| < \epsilon$. In particular, by letting $\epsilon = 1$, we see that there is $K \in \mathbb{N}$ such that $n > K$ implies $|s_n - s| < 1$. Let

$$M = \max\{|s_0 - s|, |s_1 - s|, \dots, |s_K - s|, 1\}.$$

Then if $n \in \mathbb{N}$, by the Triangle Inequality (and SACT), we have

$$|s_n| = |(s_n - s) + s| \leq |s_n - s| + |s| \leq M + |s|.$$

So, (s_n) is bounded by $M + |s|$. $\quad\square$

LEVEL 2

3. Define the sequence (s_n) recursively as follows: Let $s_0 = 2$ and for $n \in \mathbb{N}$, let $s_{n+1} = \frac{1}{3}(2s_n + 6)$. Prove that (s_n) converges and find the limit.

Proof: We have $s_1 = \frac{1}{3}(2 \cdot 2 + 6) = \frac{10}{3} \leq 6$ and $s_2 = \frac{1}{3}\left(2 \cdot \frac{10}{3} + 6\right) = \frac{1}{3}\left(\frac{20+18}{3}\right) = \frac{38}{9} \leq 6$. We now show by induction that for all $n \in \mathbb{N}$, $s_n \leq 6$. Let $k \in \mathbb{N}$ and assume that $s_n \leq 6$. Then we have $s_{n+1} = \frac{1}{3}(2s_n + 6) \leq \frac{1}{3}(2 \cdot 6 + 6) = \frac{18}{3} = 6$. By the Principle of Mathematical Induction, we have shown that (s_n) is bounded above by 6.

Next, we use the Principle of Mathematical Induction again to show that (s_n) is increasing. The base case is $s_0 = 2 \leq \frac{10}{3} = \frac{1}{3}(2 \cdot 2 + 6) = s_1$. For the inductive step, if we assume that $k \in \mathbb{N}$ and $s_k \leq s_{k+1}$, then we have $s_{k+1} = \frac{1}{3}(2s_k + 6) \leq \frac{1}{3}(2s_{k+1} + 6) = s_{k+2}$.

141

Since (s_n) is bounded above and increasing, by the Monotone Convergence Theorem, (s_n) converges to some real number L.

Now, we must have $L = \lim\limits_{n\to\infty} s_n = \lim\limits_{n\to\infty} s_{n+1} = \lim\limits_{n\to\infty}\left[\frac{1}{3}(2s_n + 6).\right] = \frac{1}{3}\left(2\lim\limits_{n\to\infty} s_n + 6\right) = \frac{1}{3}(2L + 6)$. So, $3L = 2L + 6$, and therefore, $L = 6$. $\qquad\square$

4. Let (s_n) and (t_n) be convergent sequences. Prove that $(s_n + t_n)$ converges and
$$\lim_{n\to\infty}(s_n + t_n) = \lim_{n\to\infty} s_n + \lim_{n\to\infty} t_n.$$

Proof: Suppose that $\lim\limits_{n\to\infty} s_n = s$ and $\lim\limits_{n\to\infty} t_n = t$, and let $\epsilon > 0$. Since $\lim\limits_{n\to\infty} s_n = s$, there is $K_1 \in \mathbb{N}$ such that $n > K_1 \to |s_n - s| < \frac{\epsilon}{2}$. Since $\lim\limits_{n\to\infty} t_n = t$, there is $K_2 \in \mathbb{N}$ such that $n > K_2 \to |t_n - t| < \frac{\epsilon}{2}$. Let $K = \max\{K_1, K_2\}$ and suppose that $n > K$. By the Triangle Inequality (and SACT), we have
$$|s_n + t_n - (s + t)| = |(s_n - s) + (t_n - t)| \le |s_n - s| + |t_n - t| < \frac{\epsilon}{2} + \frac{\epsilon}{2} = \epsilon.$$

So, $\lim\limits_{n\to\infty}(s_n + t_n) = s + t = \lim\limits_{n\to\infty} s_n + \lim\limits_{n\to\infty} t_n$. $\qquad\square$

5. Let (s_n) be a sequence. We say that (s_n) diverges to ∞, written $\lim\limits_{n\to\infty} s_n = \infty$, if for any $M \in \mathbb{R}$, there is $K \in \mathbb{N}$ such that $n > K$ implies $s_n > M$. Similarly, we say that (s_n) diverges to $-\infty$, written $\lim\limits_{n\to\infty} s_n = -\infty$, if for any $M \in \mathbb{R}$, there is $K \in \mathbb{N}$ such that $n > K$ implies $s_n < M$. Prove that if (s_n) is an unbounded increasing sequence, then $\lim\limits_{n\to\infty} s_n = \infty$ and if (s_n) is an unbounded decreasing sequence, then $\lim\limits_{n\to\infty} s_n = -\infty$.

Proof: Suppose that (s_n) is an unbounded increasing sequence and let $M \in \mathbb{R}$. Since (s_n) is unbounded, there is $K \in \mathbb{N}$ such that $s_K > M$. If $n > K$, then since (s_n) is increasing, $s_n > s_K$, and so, $s_n > M$. So, $\lim\limits_{n\to\infty} s_n = \infty$.

Now, suppose that (s_n) is an unbounded decreasing sequence and let $M \in \mathbb{R}$. Since (s_n) is unbounded, there is $K \in \mathbb{N}$ such that $s_K < M$. If $n > K$, then since (s_n) is decreasing, $s_n < s_K$, and so, $s_n < M$. So, $\lim\limits_{n\to\infty} s_n = -\infty$. $\qquad\square$

LEVEL 3

6. Let (s_n) be a sequence. For $k \in \mathbb{N}$, the sequence $(s_n)_{n=k}^{\infty} = (s_k, s_{k+1}, s_{k+2}, \ldots)$ is called the **k-tail** of the sequence (s_n). For example, the 0-tail of (s_n) is $(s_n)_{n=0}^{\infty} = (s_0, s_1, s_2, \ldots)$. This is just the original sequence. As another example, the 1-tail of (s_n) is $(s_n)_{n=1}^{\infty} = (s_1, s_2, s_3, \ldots)$. Let $k \in \mathbb{N}$. Prove that (s_n) converges if and only if the k-tail of (s_n) converges.

Proof: First suppose that (s_n) converges to s. For each $n \in \mathbb{N}$, let $t_n = s_{n+k}$. Then (t_n) is the k-tail of (s_n). Let $\epsilon > 0$. Since $s_n \to s$, there is $K \in \mathbb{N}$ such that $n > K$ implies $|s_n - s| < \epsilon$. Then $n > K$ implies $n + k > K$, and so, $|t_n - s| = |s_{n+k} - s| < \epsilon$. So, $t_n \to s$.

Conversely, suppose that $(t_n) = (s_{n+k}) = (s_n)_{n=k}^{\infty}$ converges to s. Let $\epsilon > 0$. Then there is $K_1 \in \mathbb{N}$ such that $n > K_1$ implies $|t_n - s| < \epsilon$. Let $K = K_1 + k$ and let $n > K$. Then $n - k > K_1$. Therefore, $|s_n - s| = |t_{n-k} - s| < \epsilon$. So, $s_n \to s$. $\square$

7. Let (s_n) and (t_n) be convergent sequences. Prove that $(s_n t_n)$ converges and
$$\lim_{n \to \infty} s_n t_n = \left(\lim_{n \to \infty} s_n \right) \left(\lim_{n \to \infty} t_n \right).$$

Proof: Suppose that $\lim_{n \to \infty} s_n = s$ and $\lim_{n \to \infty} t_n = t$, and let $\epsilon > 0$. Since $\lim_{n \to \infty} t_n = t$, there is $K_1 \in \mathbb{N}$ such that $n > K_1$ implies $|t_n - t| < 1$. Now, $|t_n - t| < 1$ is equivalent to $-1 < t_n - t < 1$, or by adding t, $t - 1 < t_n < t + 1$. Let $M = \max\{|t - 1|, |t + 1|\}$. Then, $n > K_1$ implies $-M < t_n < M$, or equivalently, $|t_n| < M$. Note also that $M > 0$. Therefore, $M + |s| > 0$.

Now, since $\lim_{n \to \infty} s_n = s$, there is $K_2 \in \mathbb{N}$ such that $n > K_2 \to |s_n - s| < \frac{\epsilon}{M+|s|}$. Since $\lim_{n \to \infty} t_n = t$, there is $K_3 \in \mathbb{N}$ such that $n > K_3 \to |t_n - t| < \frac{\epsilon}{M+|s|}$. Let $K = \max\{K_1, K_2, K_3\}$ and suppose that $n > K$.

Since $n > K_1$, $|t_n| < M$. Since $n > K_2$, $|s_n - s| < \frac{\epsilon}{M+|s|}$. Since $n > K_3$, $|t_n - t| < \frac{\epsilon}{M+|s|}$. By the Triangle Inequality (and SACT), we have

$$|s_n t_n - st| = |(s_n t_n - st_n) + (st_n - st)|$$
$$\leq |s_n t_n - st_n| + |st_n - st| = |s_n - s||t_n| + |s||t_n - t|$$
$$< \frac{\epsilon}{M+|s|} \cdot M + |s| \frac{\epsilon}{M+|s|} = \frac{\epsilon}{M+|s|}(M + |s|) = \epsilon.$$

So, $\lim_{n \to \infty} s_n t_n = st = \left(\lim_{n \to \infty} s_n \right) \left(\lim_{n \to \infty} t_n \right).$ $\square$

LEVEL 4

8. Prove that a sequence converges if and only if every one of its subsequences converges.

Proof: First suppose that (s_n) converges to s and let $\left(s_{n_k} \right)$ be a subsequence of (s_n). We will show that $\left(s_{n_k} \right)$ also converges to s. Let $\epsilon > 0$. Since (s_n) converges to s, there is $K_1 \in \mathbb{N}$ such that $n > K_1$ implies $|s_n - s| < \epsilon$. Since (n_k) is an increasing sequence, there is $K \in \mathbb{N}$ such that $k > K$ implies $n_k > K_1$. So, $k > K \to n_k > K_1 \to \left| s_{n_k} - s \right| < \epsilon$. Therefore, $\left(s_{n_k} \right)$ converges to s.

Since every sequence is a subsequence of itself, the converse is obvious. $\square$

9. Let $(x_n), (y_n), (z_n)$ be sequences such that for all $n \in \mathbb{N}$, $x_n \leq y_n \leq z_n$. Prove that if (x_n) and (z_n) both converge to the same limit L, then (y_n) also converges to L. This result is known as the **Squeeze Theorem**.

Proof: Let $\epsilon > 0$. Since $\lim_{n\to\infty} x_n = L$, there is $K_1 \in \mathbb{N}$ such that $n > K_1$ implies $|x_n - L| < \epsilon$. Since $\lim_{n\to\infty} z_n = L$, there is $K_2 \in \mathbb{N}$ such that $n > K_2$ implies $|x_n - L| < \epsilon$. Let $K = \max\{K_1, K_2\}$ and let $n > K$. Then $n > K_1$, so that $|x_n - L| < \epsilon$, or equivalently, $-\epsilon < x_n - L < \epsilon$, or $L - \epsilon < x_n < L + \epsilon$. We will need only that $L - \epsilon < x_n$. Similarly, we have $n > K_2$, so that $|z_n - L| < \epsilon$, or equivalently, $-\epsilon < z_n - L < \epsilon$, or $L - \epsilon < z_n < L + \epsilon$. We will need only that $z_n < L + \epsilon$. Now, we have $L - \epsilon < x_n \leq y_n \leq z_n < L + \epsilon$. So, $-\epsilon < y_n - L < \epsilon$, or equivalently, $|y_n - L| < \epsilon$. Since $\epsilon > 0$ was arbitrary, $\lim_{n\to\infty} y_n = L$. $\qquad\square$

> 10. Let $A \subseteq \mathbb{R}$ with $[k, \infty) \subseteq A$ for some $k \in \mathbb{N}$. Let $L \in \mathbb{R} \cup \{-\infty, \infty\}$ and let $f: A \to \mathbb{R}$ be a function such that $\lim_{x\to\infty} f(x) = L$. For each $n \in \mathbb{N}$ with $n \geq k$, let $s_n = f(n)$. Prove that $\lim_{n\to\infty} s_n = L$.

Proof: First assume that $L \in \mathbb{R}$. Let $\epsilon > 0$. Since $\lim_{x\to\infty} f(x) = L$, by the solution to part (ii) of Problem 16 in Problem Set 8, there is $K_1 > 0$ such that $x > K_1$ implies $|f(x) - L| < \epsilon$. By the Archimedean Property of $\mathbb{R}$ (Theorem 6.16), there is $K \in \mathbb{N}$ with $K > \max\{k, K_1\}$. Let $n \in \mathbb{N}$ with $n > K$. Then $n \geq k$, and so, $s_n = f(n)$. Also, $n > K_1$. So, $|s_n - L| = |f(n) - L| < \epsilon$. Since $\epsilon > 0$ was arbitrary, it follows that $\lim_{n\to\infty} s_n = L$.

Next, assume that $L = \infty$. Let $M \in \mathbb{R}$. Since $\lim_{x\to\infty} f(x) = \infty$, by the solution to part (iv) of Problem 16 in Problem Set 8, there is $K_1 > 0$ such that $x > K_1$ implies $f(x) > M$. By the Archimedean Property of $\mathbb{R}$ (Theorem 6.16), there is $K \in \mathbb{N}$ with $K > \max\{k, K_1\}$. Let $n \in \mathbb{N}$ with $n > K$. Then $n \geq k$, and so, $s_n = f(n)$. Also, $n > K_1$. So, $s_n = f(n) > M$. Since $M \in \mathbb{R}$ was arbitrary, it follows that $\lim_{n\to\infty} s_n = \infty$.

Finally, assume that $L = -\infty$. Let $M \in \mathbb{R}$. Since $\lim_{x\to\infty} f(x) = -\infty$, by the solution to part (v) of Problem 16 in Problem Set 8, there is $K_1 > 0$ such that $x > K_1$ implies $f(x) < -M$. By the Archimedean Property of $\mathbb{R}$ (Theorem 6.16), there is $K \in \mathbb{N}$ with $K > \max\{k, K_1\}$. Let $n \in \mathbb{N}$ with $n > K$. Then $n \geq k$, and so, $s_n = f(n)$. Also, $n > K_1$. So, $s_n = f(n) < -M$. Since $M \in \mathbb{R}$ was arbitrary, $\lim_{n\to\infty} s_n = \infty$. $\qquad\square$

> 11. Prove that $\lim_{n\to\infty} n^{\frac{1}{n}} = 1$.

Proof: Define $f: (0, \infty) \to \mathbb{R}$ by $f(x) = x^{\frac{1}{x}}$.

$$\lim_{x\to\infty} x^{\frac{1}{x}} = \lim_{x\to\infty} e^{\frac{1}{x}\ln x} = e^{\lim_{x\to\infty}\frac{1}{x}\ln x} = e^{\lim_{x\to\infty}\frac{\ln x}{x}} = e^{\lim_{x\to\infty}\frac{1}{x}} = e^0 = 1.$$

For the first equality, we used the definition of the general exponential function (see Lesson 12).

For the second equality, we used the continuity of e^x and $\frac{1}{x}$ on $(0, \infty)$ together with the Note following Example 12.21.

For the fourth equality, we used L'Hôpital's rule.

Finally, by Problem 10 above, $\lim_{n\to\infty} n^{\frac{1}{n}} = 1$. $\qquad\square$

12. Let (s_n) and (t_n) be convergent sequences such that $\lim\limits_{n\to\infty} t_n \neq 0$ and suppose that for all $n \in \mathbb{N}$, $t_n \neq 0$. Prove that $\left(\frac{s_n}{t_n}\right)$ converges and

$$\lim_{n\to\infty} \frac{s_n}{t_n} = \frac{\lim\limits_{n\to\infty} s_n}{\lim\limits_{n\to\infty} t_n}.$$

Proof: Suppose that $\lim\limits_{n\to\infty} s_n = s$ and $\lim\limits_{n\to\infty} t_n = t$, and let $\epsilon > 0$. Since $\lim\limits_{n\to\infty} t_n = t$, there is $K_1 \in \mathbb{N}$ such that $n < K_1$ implies $|t_n - t| < \frac{|t|}{2}$. Now, $|t_n - t| < \frac{|t|}{2}$ is equivalent to $-\frac{|t|}{2} < t_n - t < \frac{|t|}{2}$, or by adding t, $t - \frac{|t|}{2} < t_n < t + \frac{|t|}{2}$. If $t > 0$, we have $\frac{t}{2} < t_n < \frac{3t}{2}$. If $t < 0$, we have $\frac{3t}{2} < t_n < \frac{t}{2}$. In both cases, we have $\frac{|t|}{2} < |t_n| < \frac{3|t|}{2}$. Let $M = \frac{|t|}{2}$. Then $|t_n| > M$, and so, $\frac{1}{|t_n|} < \frac{1}{M}$.

Now, since $\lim\limits_{n\to\infty} s_n = s$, there is $K_2 \in \mathbb{N}$ such that $n > K_2$ implies $|s_n - s| < \frac{M|t|\epsilon}{|s|+|t|}$. Since $\lim\limits_{n\to\infty} t_n = t$, there is $K_3 > 0$ such that $n > K_3$ implies $|t_n - t| < \frac{M|t|\epsilon}{|s|+|t|}$. Let $K = \max\{K_1, K_2. K_3\}$ and suppose that $n > K$. Then since $K \geq K_1$, $\frac{1}{|t_n|} < \frac{1}{M}$. Since $K \geq K_2$, $|s_n - s| < \frac{M|t|\epsilon}{|s|+|t|}$. Since $K \geq K_3$, $|t_n - t| < \frac{M|t|\epsilon}{|s|+|t|}$. By the Triangle Inequality (and SACT), we have

$$\left|\frac{s_n}{t_n} - \frac{s}{t}\right| = \left|\frac{ts_n - st_n)}{tt_n}\right| = \left|\frac{ts_n - ts + ts - st_n}{tt_n}\right| = \left|\frac{ts_n - ts}{tt_n} + \frac{ts - st_n}{tt_n}\right|$$

$$\leq \left|\frac{ts_n - ts}{tt_n}\right| + \left|\frac{ts - st_n}{tt_n}\right| = \frac{|s_n - s|}{|t_n|} + \left|\frac{s}{t}\right|\frac{|t - t_n|}{|t_n|} = \frac{|s_n - s|}{|t_n|} + \left|\frac{s}{t}\right|\frac{|t_n - t|}{|t_n|}$$

$$= \frac{1}{|t_n|}\left(|s_n - s| + \left|\frac{s}{t}\right||t_n - t|\right) < \frac{1}{M}\left(\frac{M|t|\epsilon}{|s| + |t|} + \left|\frac{s}{t}\right|\frac{M|t|\epsilon}{|s| + |t|}\right) = \frac{1}{M}\cdot\frac{M|t|\epsilon}{|s| + |t|}\left(1 + \left|\frac{s}{t}\right|\right)$$

$$= \frac{|t|\epsilon}{|s| + |t|}\left(\frac{|t| + |s|}{|t|}\right) = \epsilon.$$

So, $\lim\limits_{n\to\infty} \frac{s_n}{t_n} = \frac{s}{t} = \frac{\lim\limits_{n\to\infty} s_n}{\lim\limits_{n\to\infty} t_n}$. $\qquad\square$

13. Let (s_n) be a bounded sequence, let $a = \lim\inf s_n$ and let $b = \lim\sup s_n$. Prove that there is a subsequence of (s_n) converging to a and another subsequence of (s_n) converging to b. Note that this provides a proof of the Bolzano-Weierstrauss Theorem (Theorem 8.27).

Proof: Let $x_n = \inf\{s_j \mid j \geq n\}$ so that $\lim\limits_{n\to\infty} x_n = \lim\inf s_n = a$. We define a subsequence of (s_n) inductively as follows: Let $n_0 = 0$. Suppose we have chosen $n_0 < n_1 < \cdots < n_k$. Choose $m > n_k$ such that $x_{n_k+1} = \inf\{s_j \mid j \geq n_k + 1\} > s_m - \frac{1}{k+1}$. Let $n_{k+1} = m$. We will show that the subsequence (s_{n_k}) converges to a.

Since $n_{k-1} + 1 \leq n_k$, we have $\{s_j \mid j \geq n_k\} \subseteq \{s_j \mid j \geq n_{k-1} + 1\}$, and so, by Problem 3 from Problem Set 11, $x_{n_k} \geq x_{n_{k-1}+1}$. Since $s_{n_k} \in \{s_j \mid j \geq n_k\}$, $s_{n_k} \geq x_{n_k}$. Therefore, if $k \geq 2$, we have

$$\left| s_{n_k} - x_{n_k} \right| = s_{n_k} - x_{n_k} \leq s_{n_k} - x_{n_{k-1}+1} < \frac{1}{k}.$$

Let $\epsilon > 0$. Since $x_n \to a$, by the proof of Problem 8 above, $x_{n_k} \to a$. So, there is $K_1 \in \mathbb{N}$ such that $k > K_1$ implies $\left| x_{n_k} - a \right| < \frac{\epsilon}{2}$. By the Archimedean Property of $\mathbb{R}$, we can find $K_2 \in \mathbb{N}$ such that $K_2 > \frac{2}{\epsilon}$, or equivalently, $\frac{1}{K_2} < \frac{\epsilon}{2}$. Let $K = \max\{K_1, K_2\}$. Then $k > K$ implies

$$\left| s_{n_k} - a \right| = \left| s_{n_k} - x_{n_k} + x_{n_k} - a \right| \leq \left| s_{n_k} - x_{n_k} \right| + \left| x_{n_k} - a \right| < \frac{1}{k} + \frac{\epsilon}{2} < \frac{1}{K_2} + \frac{\epsilon}{2} < \frac{\epsilon}{2} + \frac{\epsilon}{2} = \epsilon.$$

So, $s_{n_k} \to a$.

Next, let $x_n = \sup\{s_j \mid j \geq n\}$ so that $\lim_{n \to \infty} x_n = \limsup s_n = b$. We define a subsequence of (s_n) inductively as follows: Let $n_0 = 0$. Suppose we have chosen $n_0 < n_1 < \cdots < n_k$. Choose $m > n_k$ such that $x_{n_k+1} = \sup\{s_j \mid j \geq n_k + 1\} < s_m + \frac{1}{k+1}$. Let $n_{k+1} = m$. We will show that the subsequence (s_{n_k}) converges to b.

Since $n_{k-1} + 1 \leq n_k$, we have $\{s_j \mid j \geq n_k\} \subseteq \{s_j \mid j \geq n_{k-1} + 1\}$, and so, by Problem 3 from Problem Set 11, $x_{n_k} \leq x_{n_{k-1}+1}$. Since $s_{n_k} \in \{s_j \mid j \geq n_k\}$, $s_{n_k} \leq x_{n_k}$. Therefore, if $k \geq 2$, we have

$$\left| s_{n_k} - x_{n_k} \right| = x_{n_k} - s_{n_k} \leq x_{n_{k-1}+1} - s_{n_k} < \frac{1}{k}.$$

Let $\epsilon > 0$. Since $x_n \to b$, by the proof of Problem 8 above, $x_{n_k} \to b$. So, there is $K_1 \in \mathbb{N}$ such that $k > K_1$ implies $\left| x_{n_k} - b \right| < \frac{\epsilon}{2}$. By the Archimedean Property of $\mathbb{R}$, we can find $K_2 \in \mathbb{N}$ such that $K_2 > \frac{2}{\epsilon}$, or equivalently, $\frac{1}{K_2} < \frac{\epsilon}{2}$. Let $K = \max\{K_1, K_2\}$. Then $k > K$ implies

$$\left| s_{n_k} - b \right| = \left| s_{n_k} - x_{n_k} + x_{n_k} - b \right| \leq \left| s_{n_k} - x_{n_k} \right| + \left| x_{n_k} - b \right| < \frac{1}{k} + \frac{\epsilon}{2} < \frac{1}{K_2} + \frac{\epsilon}{2} < \frac{\epsilon}{2} + \frac{\epsilon}{2} = \epsilon.$$

So, $s_{n_k} \to b$. $\qquad \square$

14. Let (s_n) be a bounded sequence. Prove that (s_n) converges if and only if $\limsup s_n = \liminf s_n$.

Proof: First note that since (s_n) is bounded, by Theorem 14.9, $\limsup s_n$ and $\liminf s_n$ are finite real numbers.

Now, assume that $\liminf s_n = \limsup s_n$. Let $x_n = \sup\{s_k \mid k \geq n\}$ and $y_n = \inf\{s_k \mid k \geq n\}$. For all $n \in \mathbb{N}$, we have $y_n \leq s_n \leq x_n$. Also, $\lim_{n \to \infty} y_n = \liminf s_n = \limsup s_n = \lim_{n \to \infty} x_n$. By the Squeeze Theorem (Problem 9 above), $\lim_{n \to \infty} s_n$ exists and is equal to $\liminf s_n = \limsup s_n$.

Conversely, assume that (s_n) converges to s. By Problem 13 above, there is a subsequence (s_{n_k}) of (s_n) such that $s_{n_k} \to \limsup s_n$. By the proof of Problem 8 above, $s_{n_k} \to s$. By Problem 1 above, $\limsup s_n = s$. Similarly, $\liminf s_n = s$. So, $\limsup s_n = \liminf s_n$. $\qquad\square$

15. Determine with proof if the following statement is true or false: There is a sequence (x_n) such that for all $r \in \mathbb{R}$, (x_n) has a subsequence converging to r.

Proof: The statement is true. Since $\mathbb{Q}$ is countable, there is a bijection $f: \mathbb{N} \to \mathbb{Q}$. This gives us a sequence $(s_n) = (f(n))$. Let $r \in \mathbb{R}$. We will construct a subsequence (s_{n_k}) such that $s_{n_k} \to r$. For each $k \in \mathbb{N}$, we simply choose a rational number $q_k \in \left(r - \frac{1}{k}, r + \frac{1}{k}\right)$ and let $n_k \in \mathbb{N}$ with $s_{n_k} = q_k$. Let $\epsilon > 0$. By the Archimedean Property of $\mathbb{R}$, there is $K \in \mathbb{N}$ with $K > \frac{1}{\epsilon}$. If $j > K$, then we have $\left(r - \frac{1}{j}, r + \frac{1}{j}\right) \subseteq \left(r - \frac{1}{K}, r + \frac{1}{K}\right)$, and so, $s_{n_j} \in \left(r - \frac{1}{K}, r + \frac{1}{K}\right)$. Therefore, $r - \frac{1}{K} < s_{n_j} < r + \frac{1}{K}$, and so, $\left|s_{n_j} - r\right| < \frac{1}{K} < \epsilon$. So, $s_{n_j} \to r$. $\qquad\square$

Problem Set 15

LEVEL 1

1. Prove that the sum of a convergent series is unique.

Proof: Let $\sum s_n$ be a convergent series and let (x_n) be the corresponding sequence of partial sums. The sum of the series is equal to $\lim\limits_{n\to\infty} x_n$, which is unique by Problem 1 in Problem Set 14. □

2. Let $k \in \mathbb{N}$ and suppose that S and T are finite real numbers such that

$$S = \sum_{n=k}^{\infty} s_n \quad \text{and} \quad T = \sum_{n=k}^{\infty} t_n$$

Prove that if $c, d \in \mathbb{R}$, then

$$\sum_{n=k}^{\infty} (cs_n + dt_n) = cS + dT.$$

Proof: Let (x_n) be the sequence of partial sums of $\sum s_n$ and let (y_n) be the sequence of partial sums of $\sum t_n$. By Problems 4 and 7 from Problem Set 14, we have

$$cS + dT = c \lim_{n\to\infty} x_n + d \lim_{n\to\infty} y_n = \left(\lim_{n\to\infty} c\right)\left(\lim_{n\to\infty} x_n\right) + \left(\lim_{n\to\infty} d\right)\left(\lim_{n\to\infty} y_n\right)$$

$$= \lim_{n\to\infty} cx_n + \lim_{n\to\infty} dy_n = \lim_{n\to\infty} (cx_n + dy_n) = \sum_{n=k}^{\infty} (cs_n + dt_n).$$ □

LEVEL 2

3. Use the Root Test to prove that the following series converges:

$$\sum_{n=0}^{\infty} s_n, \text{where } s_n = \begin{cases} \dfrac{1}{2^n} & \text{if } n \text{ is odd.} \\ 0 & \text{if } n \text{ is even.} \end{cases}$$

Proof: We have $\limsup |s_n|^{\frac{1}{n}} = \lim\limits_{n\to\infty} \left|\frac{1}{2^n}\right|^{\frac{1}{n}} \lim\limits_{n\to\infty} \frac{1}{2} = \frac{1}{2} < 1$. So, by the Root Test, the given series converges absolutely. □

4. Suppose that we apply the Root Test to the series $\sum s_n$ and we are able to conclude that either the series converges or diverges. Will the Ratio Test necessarily provide the same information? (Compare this to Problem 14 below.)

Solution: No. Problem 3 above provides an example of a convergent series where the Root Test is successful in determining that convergence. The Ratio Test cannot be applied because if n is even, then $\frac{s_{n+1}}{s_n}$ is undefined.

148

If we replace $\frac{1}{2^n}$ by 2^n in the series in Problem 3, then we get a series that diverges by the Root Test. Once again, the Ratio Test cannot be applied because if n is even, then $\frac{s_{n+1}}{s_n}$ is undefined.

5. Prove that a conditionally convergent series contains infinitely many positive terms and infinitely many negative terms.

Proof: Let $\sum_{n=0}^{\infty} s_n$ be a conditionally convergent series. First suppose toward contradiction that the series has only finitely many negative terms. Then there is $k \in \mathbb{N}$ such that $\sum_{n=k}^{\infty} s_n$ consists of only positive terms. Therefore, $\sum_{n=k}^{\infty} s_n$ is absolutely convergent. Suppose that $\sum_{n=k}^{\infty} s_n$ converges to L. Then $\sum_{n=0}^{\infty} |s_n|$ converges to $\sum_{n=0}^{k-1} |s_n| + L$. This shows that $\sum_{n=0}^{\infty} s_n$ is absolutely convergent, which is a contradiction. So, $\sum_{n=0}^{\infty} s_n$ has infinitely many negative terms.

Now, suppose toward contradiction that the series has only finitely many positive terms. Then there is $k \in \mathbb{N}$ such that $\sum_{n=k}^{\infty} s_n$ consists of only negative terms. Suppose that $\sum_{n=k}^{\infty} s_n$ converges to L. Then by Problem 2 above, $\sum_{n=k}^{\infty} |s_n|$ converges to $-L$. So, $\sum_{n=0}^{\infty} |s_n|$ converges to $\sum_{n=0}^{k-1} |s_n| - L$. This shows that $\sum_{n=0}^{\infty} s_n$ is absolutely convergent, which is a contradiction. So, $\sum_{n=0}^{\infty} s_n$ has infinitely many positive terms. $\qquad\square$

6. Let $\sum s_n$ be a conditionally convergent series. For each $n \in \mathbb{N}$, let $s_n^+ = \frac{s_n + |s_n|}{2}$ and $s_n^- = \frac{s_n - |s_n|}{2}$. Prove that the series $\sum s_n^+$ consisting of only the positive terms from $\sum s_n$ is divergent and the series $\sum s_n^-$ consisting of only the negative terms from $\sum s_n$ is divergent.

Proof: Suppose toward contradiction that $\sum s_n^+$ converges. Then by Problem 2 above, $\sum |s_n| = \sum (2s_n^+ - s_n) = 2 \sum s_n^+ - \sum s_n$ converges. So, $\sum s_n$ is absolutely convergent, which is a contradiction. So, $\sum s_n^+$ diverges.

Similarly, suppose toward contradiction that $\sum s_n^-$ converges. Then by Problem 2 above, $\sum |s_n| = \sum (s_n - 2s_n^-) = \sum s_n - 2 \sum s_n^-$ converges. So, once again, $\sum s_n$ is absolutely convergent, which is a contradiction. Therefore, $\sum s_n^-$ diverges. $\qquad\square$

LEVEL 3

7. Prove that if a series converges absolutely, then it converges.

Proof: Let $\sum s_n$ be an absolutely convergent series. Then $\sum |s_n|$ converges. Let $\epsilon > 0$. By the Cauchy Convergence Criterion for Series (Lemma 15.8), there is $K \in \mathbb{N}$ such that $m \geq n > K$ implies

$$\sum_{k=n+1}^{m} |s_k| = \left| \sum_{k=n+1}^{m} |s_k| \right| < \epsilon.$$

By the Triangle Inequality,

$$\left| \sum_{k=n+1}^{m} s_k \right| \leq \sum_{k=n+1}^{m} |s_k| < \epsilon.$$

So, once again, by the Cauchy Convergence Criterion for Series, $\sum s_n$ converges. $\square$

8. Let (s_n) be a sequence with $s_n \neq 0$ for all $n \in \mathbb{N}$ and suppose that there is $r \in (0,1)$ and $K \in \mathbb{N}$ such that $n > K \rightarrow \left|\frac{s_{n+1}}{s_n}\right| < r$. Prove that $\sum s_n$ converges absolutely. This is Lemma 15.22.

Proof: We first prove by induction that for all $n \in \mathbb{N}$ with $n \geq 2$, $|s_{K+n}| < |s_{K+1}|r^{n-1}$.

Base Case ($t = 2$): Since $K + 1 > K$, by assumption, $|s_{K+2}| < |s_{K+1}|r = |s_{K+1}|r^1 = |s_{K+1}|r^{2-1}$.

Inductive step: Assume that $t \geq 2$ and $|s_{K+t}| < |s_{K+1}|r^{t-1}$. Then we have

$$\left|s_{K+(t+1)}\right| < |s_{K+t}|r < |s_{K+1}|r^{t-1}r = |s_{K+1}|r^{(t+1)-1}.$$

By the Principle of Mathematical Induction, for all $n \in \mathbb{N}$ with $n \geq 2$, $|s_{K+n}| < |s_{K+1}|r^{n-1}$.

Since $\sum r^{n-1}$ is a convergent geometric series, by Problem 2 above, $\sum |s_{K+1}|r^{n-1}$ is a convergent series. So, by the Comparison Test (Theorem 15.13), $\sum |s_{K+n}|$ converges. So, $\sum s_n$ converges absolutely. $\square$

9. Let (s_n) be a sequence with $s_n \neq 0$ for all $n \in \mathbb{N}$ and suppose that there is $K \in \mathbb{N}$ such that $n > K \rightarrow \left|\frac{s_{n+1}}{s_n}\right| > 1$. Prove that $\sum s_n$ diverges. This is Lemma 15.23.

Proof: We first prove by induction that for all $n \in \mathbb{N}$ with $n \geq 2$, $|s_{K+n}| > |s_{K+1}|$.

Base Case ($t = 2$): Since $K + 1 > K$, by assumption, $|s_{K+2}| > |s_{K+1}|$.

Inductive step: Assume that $t \geq 2$ and $|s_{K+t}| > |s_{K+1}|$. Then we have

$$\left|s_{K+(t+1)}\right| = \left|s_{(K+t)+1}\right| > |s_{K+t}| > |s_{K+1}|.$$

By the Principle of Mathematical Induction, for all $n \in \mathbb{N}$ with $n \geq 2$, $|s_{K+n}| > |s_{K+1}|$.

Since $\lim\limits_{n\to\infty} |s_{K+1}| = |s_{K+1}| > 0$, it follows that $\lim\limits_{n\to\infty} |s_{K+1}| > 0$. So, $\lim\limits_{n\to\infty} s_{K+1} \neq 0$. Therefore, by the Divergence Test (see part 2 of Example 15.1), $\sum s_n$ diverges. $\square$

10. Let (s_n) be a sequence with $s_n \neq 0$ for all $n \in \mathbb{N}$ and let

$$L = \limsup |s_n|^{\frac{1}{n}}$$

Prove each of the following:

$$\text{if } L < 1, \text{then} \sum_{n=0}^{\infty} s_n \text{ converges absolutely.}$$

$$\text{if } L > 1, \text{then} \sum_{n=0}^{\infty} s_n \text{ diverges.}$$

This is the **Root Test** (Theorem 15.25).

Proof: We will first prove two lemmas:

Lemma 1: Let (s_n) be a sequence and suppose that there is $r \in \mathbb{R}$ with $r < 1$ and $K \in \mathbb{N}$ such that $n > K \rightarrow |s_n|^{\frac{1}{n}} < r$. Prove that $\sum s_n$ converges absolutely.

Proof of Lemma 1: By assumption, if $n > K$, then $|s_n| < r^n$. Since $\sum r^n$ is a convergent geometric series, by the Comparison Test (Theorem 15.13), $\sum |s_n|$ converges. So, $\sum s_n$ converges absolutely. □

Lemma 2: Let (s_n) be a sequence and suppose that there is $K \in \mathbb{N}$ such that $n > K \rightarrow |s_n|^{\frac{1}{n}} > 1$. Prove that $\sum s_n$ diverges.

Proof of Lemma 2: By assumption, if $n > K$, then $|s_n| > 1^n = 1$. Since $\lim_{n \to \infty} 1 = 1$, $\lim_{n \to \infty} |s_n| > 1$. So, $\lim_{n \to \infty} s_n \neq 0$. Therefore, by the Divergence Test (see part 2 of Example 15.1), $\sum s_n$ diverges. □

Proof of Root Test: First suppose that $L < 1$. Let r be any real number such that $L < r < 1$. Then $0 < r - L < 1 - L < 1$ (because $L \geq 0$). Since $\limsup |s_n|^{\frac{1}{n}} = L$ and $r > L$, there is $K \in \mathbb{N}$ such that

$$n > K \rightarrow |s_n|^{\frac{1}{n}} < r.$$

By Lemma 1 above, $\sum s_n$ converges absolutely.

Now, suppose that $L > 1$. Since $\limsup |s_n|^{\frac{1}{n}} = L$, there is $K \in \mathbb{N}$ such that

$$n > K \rightarrow |s_n|^{\frac{1}{n}} > 1,$$

By Lemma 2 above, $\sum s_n$ diverges. □

11. Consider the series $\sum_{n=k}^{\infty} s_n$. Let (n_j) be an increasing sequence of integers with $n_0 = k$. For each $j \in \mathbb{N}^+$, let

$$t_j = \sum_{n=n_j}^{n_{j+1}-1} s_n = s_{n_j} + s_{n_j+1} + \cdots + s_{n_{j+1}-1}.$$

The series $\sum_{j=0}^{\infty} t_n$ is called a **regrouping** of $\sum_{n=k}^{\infty} s_n$.

Prove that if $\sum_{n=k}^{\infty} s_n$ is a convergent series, then any regrouping of $\sum_{n=k}^{\infty} s_n$ converges to the same value. Is there an analogous result for divergent series?

Proof: Let x_i be the ith partial sum of $\sum_{n=k}^{\infty} s_n$ and let y_i be the ith partial sum of $\sum_{j=0}^{\infty} t_n$. Then

$$y_i = \sum_{j=0}^{i} t_j = t_0 + t_1 + \cdots + t_i$$

$$= \left(s_k + s_{k+1} + \cdots + s_{n_1-1}\right) + \left(s_{n_1} + s_{n_1+1} + \cdots + s_{n_2-1}\right) + \cdots + \left(s_{n_i} + s_{n_i+1} + \cdots + s_{n_{i+1}-1}\right)$$

$$= \sum_{n=k}^{n_{i+1}-1} s_n = x_{n_{i+1}-1}.$$

So, (y_i) is a subsequence of (x_i).

If $\sum_{n=k}^{\infty} s_n$ converges to L, then (x_i) converges to L. So, by the solution to Problem 8 from Problem Set 14, (y_i) converges to L. Therefore, $\sum_{j=0}^{\infty} t_j$ converges to L.

If $\sum_{n=k}^{\infty} s_n$ diverges, then (x_i) diverges. So, by the solution to Problem 8 from Problem Set 14, (y_i) diverges. Therefore, $\sum_{j=0}^{\infty} t_j$ diverges. $\qquad\square$

LEVEL 4

12. Prove that for each $k \in \mathbb{N}^+$,

$$\sum_{n=1}^{k} \cos n = \frac{\sin\left(k + \frac{1}{2}\right) - \sin\frac{1}{2}}{2\sin\frac{1}{2}}.$$

Proof: We first show that for all $x, y \in \mathbb{R}$, $2\sin x \cos y = \sin(x + y) + \sin(x - y)$.

By part 5 of Problem 13 from Problem Set 3, we have $\sin(x + y) = \sin x \cos y + \cos x \sin y$.

By Note 2 following Example 3.24, we have

$$\sin(x - y) = \sin\big(x + (-y)\big) = \sin x \cos(-y) + \cos x \sin(-y) = \sin x \cos y - \cos x \sin y.$$

Adding these two equations yields the desired result.

Next, we prove by induction on $k \in \mathbb{N}^+$ that

$$\sum_{n=1}^{k}\left[\sin\left(\frac{1}{2} + n\right) + \sin\left(\frac{1}{2} - n\right)\right] = \sin\left(k + \frac{1}{2}\right) - \sin\frac{1}{2}.$$

Base case ($k = 1$): We have

$$\sum_{n=1}^{1}\left[\sin\left(\frac{1}{2} + n\right) + \sin\left(\frac{1}{2} - n\right)\right] = \sin\left(\frac{1}{2} + 1\right) + \sin\left(\frac{1}{2} - 1\right)$$

$$= \sin\left(1 + \frac{1}{2}\right) + \sin\left(-\frac{1}{2}\right) = \sin\left(1 + \frac{1}{2}\right) - \sin\frac{1}{2}.$$

For the last equality, we used Note 2 following Example 3.24

Inductive step: Assume that

$$\sum_{n=1}^{k}\left[\sin\left(\frac{1}{2} + n\right) + \sin\left(\frac{1}{2} - n\right)\right] = \sin\left(k + \frac{1}{2}\right) - \sin\frac{1}{2}$$

Then

$$\sum_{n=1}^{k+1}\left[\sin\left(\frac{1}{2}+n\right)+\sin\left(\frac{1}{2}-n\right)\right]$$

$$=\sum_{n=1}^{k}\left[\sin\left(\frac{1}{2}+n\right)+\sin\left(\frac{1}{2}-n\right)\right]+\sin\left(\frac{1}{2}+(k+1)\right)+\sin\left(\frac{1}{2}-(k+1)\right)$$

$$=\left(\sin\left(k+\frac{1}{2}\right)-\sin\frac{1}{2}\right)+\sin\left(\frac{1}{2}+(k+1)\right)+\sin\left(\frac{1}{2}-(k+1)\right)$$

$$=\sin\left(k+\frac{1}{2}\right)-\sin\frac{1}{2}+\sin\left(\frac{1}{2}+(k+1)\right)+\sin\left(-k-\frac{1}{2}\right)$$

$$=\sin\left((k+1)+\frac{1}{2}\right)-\sin\frac{1}{2}+\sin\left(k+\frac{1}{2}\right)-\sin\left(k+\frac{1}{2}\right)$$

$$=\sin\left((k+1)+\frac{1}{2}\right)-\sin\frac{1}{2}$$

For the fourth equality, we again used Note 2 following Example 3.24

By the Principle of Mathematical Induction, for all $k \in \mathbb{N}^+$, we have

$$\sum_{n=1}^{k}\left[\sin\left(\frac{1}{2}+n\right)+\sin\left(\frac{1}{2}-n\right)\right]=\sin\left(k+\frac{1}{2}\right)-\sin\frac{1}{2}.$$

So, we have

$$2\sin\frac{1}{2}\sum_{n=1}^{k}\cos n=\sum_{n=1}^{k}2\sin\frac{1}{2}\cos n=\sum_{n=1}^{k}\left[\sin\left(\frac{1}{2}+n\right)+\sin\left(\frac{1}{2}-n\right)\right]=\sin\left(k+\frac{1}{2}\right)-\sin\frac{1}{2}.$$

Therefore,

$$\sum_{n=1}^{k}\cos n=\frac{\sin\left(k+\frac{1}{2}\right)-\sin\frac{1}{2}}{2\sin\frac{1}{2}}\qquad\qquad\square$$

13. Let (s_n) be a bounded monotone sequence and let (t_n) be another sequence whose sequence of partial sums converges. Then the following series converges:

$$\sum_{n=1}^{k}s_n t_n$$

This is **Abel's Test** (Theorem 15.11).

Proof: Let (y_n) the sequence of partial sums of (t_n). Since (y_n) converges, by Problem 2 from Problem Set 14, (y_n) is bounded

First assume that (s_n) is a decreasing sequence. By the Monotone Convergence Theorem (Theorem 14.5), there is $s \in \mathbb{R}$ such that $s_n \to s$. For each $n \in \mathbb{N}$, let $x_n = s_n - s$. Then (x_n) is a decreasing sequence such that $x_n \to 0$. For each $n \in \mathbb{N}$, $s_n = x_n + s$, and so, $s_n t_n = x_n t_n + s t_n$. By Dirichlet's Test (Theorem 15.7), $\sum x_n t_n$ converges. By Problem 2 above, $\sum s t_n$ converges. So, again by Problem 2 above, $\sum s_n t_n = \sum x_n t_n + \sum s t_n$ converges.

Next, assume that (s_n) is an increasing sequence. For each $n \in \mathbb{N}$, let $z_n = s - s_n$. Then (z_n) is a decreasing sequence such that $z_n \to 0$. For each $n \in \mathbb{N}$, $s_n = s - z_n$, and so, $s_n t_n = s t_n - z_n t_n$. By Problem 2 above, $\sum s t_n$ converges. By Dirichlet's Test (Theorem 15.7), $\sum z_n t_n$ converges. So, by Problem 2 above, $\sum s_n t_n = \sum s t_n + \sum z_n t_n$ converges. $\qquad \square$

LEVEL 5

14. Suppose that we apply the Ratio Test to the series $\sum s_n$ and we are able to conclude that either the series converges or diverges. Prove that the Root Test will provide the same information.

Proof: Suppose that there is $L \in [0, \infty)$ such that

$$\lim_{n \to \infty} \left| \frac{s_{n+1}}{s_n} \right| = L.$$

We will show that $\lim_{n \to \infty} |s_n|^{\frac{1}{n}}$ is also equal to L, and therefore, $\limsup |s_n|^{\frac{1}{n}} = L$, proving the result.

Let $\epsilon > 0$. Then there is $K \in \mathbb{N}$ such that $n \geq K$ implies $\left| \left| \frac{s_{n+1}}{s_n} \right| - L \right| < \epsilon$, or equivalently,

$$L - \epsilon < \left| \frac{s_{n+1}}{s_n} \right| < L + \epsilon.$$

Let $n \geq K$. Then

$$|s_n| = \left| \frac{s_n}{s_{n-1}} \cdot \frac{s_{n-1}}{s_{n-2}} \cdots \frac{s_{K+1}}{s_K} \cdot s_K \right| = \left| \frac{s_n}{s_{n-1}} \right| \cdot \left| \frac{s_{n-1}}{s_{n-2}} \right| \cdots \left| \frac{s_{K+1}}{s_K} \right| \cdot |s_K|$$

By the fence-post formula (see Note 4 following Example 1.7), we have

$$(L - \epsilon)^{n-K} |s_K| < \left| \frac{s_n}{s_{n-1}} \right| \cdot \left| \frac{s_{n-1}}{s_{n-2}} \right| \cdots \left| \frac{s_{K+1}}{s_K} \right| \cdot |s_K| < (L + \epsilon)^{n-K} |s_K|.$$

Therefore,

$$(L - \epsilon)^{n-K} |s_K| < |s_n| < (L + \epsilon)^{n-K} |s_K|.$$

So, we have

$$(L - \epsilon)^{1 - \frac{K}{n}} |s_K|^{\frac{1}{n}} < |s_n|^{\frac{1}{n}} < (L + \epsilon)^{1 - \frac{K}{n}} |s_K|^{\frac{1}{n}}.$$

154

If we let $y = (L + \epsilon)^{1-\frac{K}{n}}|s_K|^{\frac{1}{n}}$, then $\ln y = \left(1 - \frac{K}{n}\right)\ln(L + \epsilon) + \frac{1}{n}\ln|s_K|$.

So, $\lim\limits_{n\to\infty} \ln y = \ln(L + \epsilon) \lim\limits_{n\to\infty}\left(1 - \frac{K}{n}\right) + \ln|s_K| \lim\limits_{n\to\infty}\frac{1}{n} = \ln(L + \epsilon) + 0 = \ln(L + \epsilon)$.

Therefore, $\lim\limits_{n\to\infty}(L + \epsilon)^{1-\frac{K}{n}}|s_K|^{\frac{1}{n}} = \lim\limits_{n\to\infty} y = \lim\limits_{n\to\infty} e^{\ln y} = e^{\lim\limits_{n\to\infty}\ln y} = e^{\ln(L+\epsilon)} = L + \epsilon$.

Similarly, $\lim\limits_{n\to\infty}(L - \epsilon)^{1-\frac{K}{n}}|s_K|^{\frac{1}{n}} = L - \epsilon$.

So, $L - \epsilon \leq \lim\limits_{n\to\infty}|s_n|^{\frac{1}{n}} \leq L + \epsilon$.

Since $\epsilon > 0$ was arbitrary, by Problem 15 from problem Set 6, $\lim\limits_{n\to\infty}|s_n|^{\frac{1}{n}} = L$, as desired. $\quad\square$

15. Prove that the alternating harmonic series converges to $\ln 2$. That is

$$\sum_{n=1}^{\infty}\frac{(-1)^{n+1}}{n} = 1 - \frac{1}{2} + \frac{1}{3} - \frac{1}{4} + \cdots = \ln 2.$$

Proof: By induction on $n \in \mathbb{N}^+$ we can prove the following (See the Note below):

$$\sum_{j=1}^{2n}\frac{(-1)^{j+1}}{j} = \sum_{j=1}^{n}\frac{1}{n+j}.$$

So, for all $n \in \mathbb{N}^+$, we have

$$\sum_{j=1}^{2n}\frac{(-1)^{j+1}}{j} = \sum_{j=1}^{n}\frac{1}{n+j} = \frac{1}{n}\sum_{j=1}^{n}\frac{1}{1+\frac{j}{n}}.$$

Therefore,

$$\sum_{j=1}^{\infty}\frac{(-1)^{j+1}}{j} = \lim_{n\to\infty}\sum_{j=1}^{2n}\frac{(-1)^{j+1}}{j} = \lim_{n\to\infty}\frac{1}{n}\sum_{j=1}^{n}\frac{1}{1+\frac{j}{n}} = \int_0^1 \frac{1}{1+x}dx = \ln|1+1| - \ln|1+0| = \ln 2.\square$$

Note: Let's use induction to prove that for all $n \in \mathbb{N}^+$,

$$\sum_{j=1}^{2n}\frac{(-1)^{j+1}}{j} = \sum_{j=1}^{n}\frac{1}{n+j}.$$

Base case ($k = 1$): We have
$$\sum_{j=1}^{2\cdot1}\frac{(-1)^{j+1}}{j} = \sum_{j=1}^{2}\frac{(-1)^{j+1}}{j} = 1 - \frac{1}{2} = \frac{1}{2} = \frac{1}{1+1} = \sum_{j=1}^{1}\frac{1}{1+j}.$$

Inductive step: Assume that for some $k \in \mathbb{N}^+$, we have

$$\sum_{j=1}^{2k} \frac{(-1)^{j+1}}{j} = \sum_{j=1}^{k} \frac{1}{k+j}.$$

Then, we have

$$\sum_{j=1}^{2(k+1)} \frac{(-1)^{j+1}}{j} = \sum_{j=1}^{2k+2} \frac{(-1)^{j+1}}{j} = \sum_{j=1}^{2k} \frac{(-1)^{j+1}}{j} + \frac{(-1)^{2k+2}}{2k+1} + \frac{(-1)^{2k+3}}{2k+2}$$

$$= \sum_{j=1}^{k} \frac{1}{k+j} + \frac{1}{2k+1} - \frac{1}{2k+2} = \frac{1}{k+1} + \sum_{j=2}^{k} \frac{1}{k+j} + \frac{1}{2k+1} - \frac{1}{2k+2}$$

$$= \frac{1}{k+1} + \sum_{j=1}^{k-1} \frac{1}{k+(j+1)} + \frac{1}{2k+1} - \frac{1}{2k+2} = \frac{1}{k+1} + \sum_{j=1}^{k} \frac{1}{(k+1)+j} - \frac{1}{2k+2}$$

$$= \sum_{j=1}^{k} \frac{1}{(k+1)+j} + \frac{1}{k+1} - \frac{1}{2k+2} = \sum_{j=1}^{k} \frac{1}{(k+1)+j} + \frac{2}{2k+2} - \frac{1}{2k+2}$$

$$= \sum_{j=1}^{k} \frac{1}{(k+1)+j} + \frac{2-1}{2k+2} = \sum_{j=1}^{k} \frac{1}{(k+1)+j} + \frac{1}{2k+2} = \sum_{j=1}^{k+1} \frac{1}{(k+1)+j}.$$

By the Principle of Mathematical Induction, for all $n \in \mathbb{N}^+$, we have

$$\sum_{j=1}^{2n} \frac{(-1)^{j+1}}{j} = \sum_{j=1}^{n} \frac{1}{n+j}. \qquad \square$$

16. Let (s_n) and (t_n) be sequences and suppose that there is $K \in \mathbb{N}$ such that $n > K$ implies that $s_n \geq 0$ and $t_n > 0$. Also, suppose that there is $L \in \mathbb{R} \cup \{\infty\}$ such that

$$\lim_{n \to \infty} \frac{s_n}{t_n} = L.$$

Prove each of the following:

$$\text{if } L \in \mathbb{R}^+, \text{then } \sum_{n=0}^{\infty} s_n \text{ converges if and only if } \sum_{n=0}^{\infty} t_n \text{ converges.}$$

$$\text{if } L = 0 \text{ and } \sum_{n=0}^{\infty} t_n \text{ converges, then } \sum_{n=0}^{\infty} s_n \text{ converges.}$$

$$\text{if } L = \infty \text{ and } \sum_{n=0}^{\infty} t_n \text{ diverges, then } \sum_{n=0}^{\infty} s_n \text{ diverges.}$$

This is the **Limit Comparison Test** (Theorem 15.15).

Proof: First assume that $L \in \mathbb{R}^+$. Since $\lim\limits_{n \to \infty} \frac{s_n}{t_n} = L$, there is $K_1 \in \mathbb{N}$ such that

$$n > K_1 \to \left| \frac{s_n}{t_n} - L \right| < \frac{L}{2},$$

or equivalently,

$$n > K_1 \to 0 < \frac{L}{2} < \frac{s_n}{t_n} < \frac{3L}{2}.$$

Let $K^* = \max\{K, K_1\}$. Since $t_n > 0$ for all $n > K$, we have $n > K^* \to \frac{L}{2} t_n < s_n < \frac{3L}{2} t_n$.

If $\sum s_n \downarrow$, then by the Comparison Test (Theorem 15.13) and the inequality $\frac{L}{2} t_n < s_n$, we have that $\sum t_n \downarrow$. Similarly, if $\sum t_n \downarrow$, then by the Comparison Test and the inequality $s_n < \frac{3L}{2} t_n$, we have that $\sum s_n \downarrow$.

Next, assume that $L = 0$. Since $\lim\limits_{n \to \infty} \frac{s_n}{t_n} = 0$, there is $K_1 \in \mathbb{N}$ such that

$$n > K_1 \to \left| \frac{s_n}{t_n} \right| < 1.$$

Let $K^* = \max\{K, K_1\}$. Since $t_n > 0$ for all $n > K$, we have $n > K^* \to 0 \leq s_n < t_n$.

If $\sum t_n \downarrow$, then by the Comparison Test, we have that $\sum s_n \downarrow$.

Finally, assume that $L = \infty$. Since $\lim\limits_{n \to \infty} \frac{s_n}{t_n} = \infty$, there is $K_1 \in \mathbb{N}$ such that

$$n > K_1 \to 0 < s_n \text{ and } \frac{s_n}{t_n} \geq 1.$$

Let $K^* = \max\{K, K_1\}$. Since $s_n, t_n > 0$ for all $n > K^*$, we have $n > K^* \to s_n \geq t_n$.

If $\sum t_n \uparrow$, then by the Comparison Test and the inequality $s_n \geq t_n$, we have that $\sum s_n \uparrow$. $\qquad \square$

Problem Set 16

LEVEL 1

1. Let $f, g: A \to \mathbb{R}$ be bounded on A. Prove that $|fg|_A \leq |f|_A \cdot |g|_A$.

Proof: Since f is bounded on A, there is $M_1 \in \mathbb{R}^+$ such that $|f|_A = M_1$. Similarly, there is $M_2 \in \mathbb{R}^+$ such that $|g|_A = M_2$. For each $x \in \mathbb{R}$, we have

$$|f(x) \cdot g(x)| = |f(x)| \cdot |g(x)| \leq M_1 M_2.$$

So, $M_1 M_2$ is an upper bound of $\{|f(x) \cdot g(x)| \mid x \in A\}$. Since $|fg|_A$ is the **least** upper bound of $\{|f(x) \cdot g(x)| \mid x \in A\}$, it follows that $|fg|_A \leq M_1 M_2 = |f|_A \cdot |g|_A$. □

2. Let $f, g: A \to \mathbb{R}$ be bounded on A. Prove that $|f + g|_A \leq |f|_A + |g|_A$.

Proof: Since f is bounded on A, there is $M_1 \in \mathbb{R}^+$ such that $|f|_A = M_1$. Similarly, there is $M_2 \in \mathbb{R}^+$ such that $|g|_A = M_2$. For each $x \in \mathbb{R}$, by the Triangle Inequality (Theorem 7.4), we have

$$|f(x) + g(x)| \leq |f(x)| + |g(x)| \leq M_1 + M_2.$$

So, $M_1 + M_2$ is an upper bound of $\{|f(x) + g(x)| \mid x \in A\}$. Since $|f + g|_A$ is the **least** upper bound of $\{|f(x) + g(x)| \mid x \in A\}$, it follows that $|f + g|_A \leq M_1 + M_2 = |f|_A + |g|_A$. □

LEVEL 2

3. A sequence of functions (f_n) is said to be **decreasing** on A if for all $x \in A$ and $n \in \mathbb{N}$, $f_n(x) \geq f_{n+1}(x)$. Find a decreasing sequence (f_n) of continuous functions on $[0, 1)$ that converges to a continuous function $f: [0, 1) \to \mathbb{R}$, but the convergence is **not** uniform on $[0, 1)$.

Solution: For each $n \in \mathbb{N}$, let $f_n: [0, 1) \to \mathbb{R}$ be defined by $f_n(x) = x^n$. If $0 \leq x < 1$, then for all $n \in \mathbb{N}$, $x^n \geq x^{n+1}$. So, (f_n) is decreasing. Since each f_n is a polynomial function, each f_n is continuous. In part 1 of Example 16.1, we saw that (f_n) converges pointwise to 0 (the constant zero function).

Now, as in part 1 of Example 16.3, let $\epsilon = \frac{1}{2}$ and for each $k \in \mathbb{N}$, let $n_k = k$ and $x_k = \left(\frac{1}{2}\right)^{\frac{1}{k}}$. Observe that since $0 < \frac{1}{2} < 1$, for each $k \in \mathbb{N}$, we have $0^{\frac{1}{k}} < \left(\frac{1}{2}\right)^{\frac{1}{k}} < 1^{\frac{1}{k}}$, or $0 < \left(\frac{1}{2}\right)^{\frac{1}{k}} < 1$. So, for each $k \in \mathbb{N}$, $x_k \in (0, 1) \subseteq [0, 1)$. Then for each $k \in \mathbb{N}$, we have

$$\left|f_{n_k}(x_k) - f(x_k)\right| = \left|f_k\left(\left(\frac{1}{2}\right)^{\frac{1}{k}}\right) - f\left(\left(\frac{1}{2}\right)^{\frac{1}{k}}\right)\right| = \left|\left(\left(\frac{1}{2}\right)^{\frac{1}{k}}\right)^k - 0\right| = \left|\frac{1}{2}\right| = \frac{1}{2}.$$

By Note 6 preceding Example 16.3, (f_n) does **not** converge uniformly on $[0, 1)$. □

4. Determine the radius of convergence and the interval of convergence of each of the following power series:

$$\text{(i)} \sum_{n=1}^{\infty} \frac{(-1)^n x^{2n}}{2^{2n}(n!)^2} \qquad \text{(ii)} \sum_{n=1}^{\infty} \left(\frac{n+1}{n}\right)^n x^n \qquad \text{(iii)} \sum_{n=1}^{\infty} \frac{5^n(x+6)^n}{\sqrt{n}}$$

Solutions:

(i) If we let $s_n = \frac{(-1)^n x^{2n}}{2^{2n}(n!)^2}$, then we have

$$\lim_{n\to\infty} \left|\frac{s_{n+1}}{s_n}\right| = \lim_{n\to\infty} \left|\frac{\dfrac{(-1)^{n+1} x^{2n+2}}{2^{2n+2}((n+1)!)^2}}{\dfrac{(-1)^n x^{2n}}{2^{2n}(n!)^2}}\right| = \lim_{n\to\infty} \left|\frac{x^2}{2^2(n+1)^2}\right| = 0.$$

Since $0 < 1$, by the Ratio Test (Theorem 15.21), the series is absolutely convergent for all $x \in \mathbb{R}$. So, the radius of convergence is ∞ and the interval of convergence is $(-\infty, \infty)$.

(ii) If we let $s_n = \left(\frac{n+1}{n}\right)^n x^n$, then we have

$$\lim \sup |s_n|^{\frac{1}{n}} = \lim \sup \left|\left(\frac{n+1}{n}\right)^n x^n\right|^{\frac{1}{n}} = \lim \sup \left|\frac{n+1}{n} \cdot x\right| = |x| \lim \sup \left(\frac{n+1}{n}\right) = |x|.$$

By the Root Test (Theorem 15.25), if $|x| < 1$, the series is absolutely convergent and if $|x| > 1$, the series is divergent. So, the radius of convergence is 1. Since $\left(\frac{n+1}{n}\right)^n = \left(1 + \frac{1}{n}\right)^n \to e$, the Divergence Test (see part 2 of Example 15.1) tells us that for $x = 1$ and $x = -1$, the series diverges. It follows that the interval of convergence is $(-1, 1)$.

(iii) If we let $s_n = \frac{5^n(x+6)^n}{\sqrt{n}}$, then we have

$$\lim_{n\to\infty} \left|\frac{s_{n+1}}{s_n}\right| = \lim_{n\to\infty} \left|\frac{\dfrac{5^{n+1}(x+6)^{n+1}}{\sqrt{n+1}}}{\dfrac{5^n(x+6)^n}{\sqrt{n}}}\right| = \lim_{n\to\infty} \left|\frac{5(x+6)\sqrt{n}}{\sqrt{n+1}}\right| = 5|x+6| \lim_{n\to\infty} \frac{\sqrt{n}}{\sqrt{n+1}} = 5|x+6|.$$

By the Ratio Test (Theorem 15.21), if $5|x+6| < 1$, or equivalently, $|x+6| < \frac{1}{5}$, the series is absolutely convergent and if $|x+6| > \frac{1}{5}$, , the series is divergent.

So, the radius of convergence is $\frac{1}{5}$.

Now, $|x+6| < \frac{1}{5}$ if and only if $-\frac{1}{5} < x+6 < \frac{1}{5}$ if and only if $-\frac{31}{5} < x < -\frac{29}{5}$.

We still need to check the endpoints. When $x = -\frac{31}{5}$, we get the convergent alternating series $\sum_{n=1}^{\infty} \frac{(-1)^n}{\sqrt{n}}$. When $x = -\frac{29}{5}$, we get the divergent p-series $\sum_{n=1}^{\infty} \frac{1}{\sqrt{n}}$. So, the interval of convergence is $\left[-\frac{31}{5}, -\frac{29}{5}\right)$.

5. For each $n \in \mathbb{N}^+$, let $f_n: [0, \infty) \to \mathbb{R}$ be defined by $f_n(x) = \frac{x}{n+x}$. Prove each of the following:

   (i) f_n converges to 0 pointwise on $[0, \infty)$.

   (ii) f_n does **not** converge uniformly on $[0, \infty)$ to 0.

   (iii) For each $c \in \mathbb{R}^+$, f_n converges uniformly on $[0, c]$ to 0.

Proofs:

(i) Let $x \in [0, \infty)$ and let $\epsilon > 0$. By the Archimedean Property of $\mathbb{R}$ (Theorem 6.16), we can choose $K \in \mathbb{N}^+$ with $K > \frac{1}{\epsilon}x - x$. Suppose that $n > K$. Then we have $n > \frac{x}{\epsilon} - x$, and so, $n + x > \frac{x}{\epsilon}$. Therefore, $\epsilon > \frac{x}{n+x}$, and so, we have

$$|f_n(x) - 0| = \frac{x}{n+x} < \epsilon.$$

Since $\epsilon > 0$ was arbitrary, $f_n(x) \to 0$.

Since $x \in [0, \infty)$ was arbitrary, (f_n) converges pointwise to 0 on $[0, \infty)$. □

(ii) For each $n \in \mathbb{N}^+$, we have

$$|f_n(n) - 0| = \frac{n}{n+n} = \frac{n}{2n} = \frac{1}{2} \geq \frac{1}{2}.$$

By Note 6 preceding Example 16.3, (f_n) does **not** converge uniformly on $[0, \infty)$ to 0. □

(iii) First observe that for each $n \in \mathbb{N}^+$, $f_n(x)$ is a strictly increasing function on $[0, c]$. To see this, we differentiate using the quotient rule (Theorem 10.18):

$$f_n'(x) = \frac{(n+x)(1) - x(1)}{(n+x)^2} = \frac{n}{(n+x)^2}$$

Since the right-hand side of the above equation is positive for all $x \in [0, c]$, the result follows from part 1 of Corollary 10.26.

So, for each $n \in \mathbb{N}^+$ and each $x \in [0, c]$, we have $f_n(x) \leq f_n(c)$.

Let $\epsilon > 0$. By the Archimedean Property of $\mathbb{R}$ (Theorem 6.16), we can choose $K \in \mathbb{N}^+$ with $K > \frac{1}{\epsilon}c - c$. Suppose that $x \in [0, c]$ and $n > K$. Then we have $n > \frac{c}{\epsilon} - c$, and so, $n + c > \frac{c}{\epsilon}$. Therefore, $\epsilon > \frac{c}{n+c}$, and so, we have

$$|f_n(x) - 0| = \frac{x}{n+x} = f_n(x) \leq f_n(c) = \frac{c}{n+c} < \epsilon.$$

Since $\epsilon > 0$ was arbitrary, f_n converges uniformly on $[0, c]$ to 0. □

6. Let (f_n) and (g_n) be sequences of functions on $A \subseteq \mathbb{R}$, let $B \subseteq A$, and suppose that (f_n) and (g_n) converge uniformly on B to f and g, respectively. Determine if each of the following statements is true or false. If true, provide a proof. If false, provide a counterexample.

    (i) $(f_n + g_n)$ converges uniformly on B to $f + g$.

    (ii) $(f_n g_n)$ converges uniformly on B to fg.

Solutions:

(i)   This is **True**.

Let $\epsilon > 0$. Since (f_n) converges uniformly on B to f, there is $K_1 \in \mathbb{N}$ such that for all $x \in B$ and $n > K_1$, $|f_n(x) - f(x)| < \frac{\epsilon}{2}$. Since (g_n) converges uniformly on B to g, there is $K_2 \in \mathbb{N}$ such that for all $x \in B$ and $n > K_2$, $|g_n(x) - g(x)| < \frac{\epsilon}{2}$. Let $K = \max\{K_1, K_2\}$. Let $x \in B$ and $n > K$. By the Triangle Inequality (Theorem 7.4),

$$|(f_n + g_n)(x) - (f + g)(x)| = |f_n(x) + g_n(x) - f(x) - g(x)|$$

$$= |(f_n(x) - f(x)) + (g_n(x) = g(x))| \le |f_n(x) - f(x)| + |g_n(x) = g(x)| < \frac{\epsilon}{2} + \frac{\epsilon}{2} = \epsilon.$$

So $(f_n + g_n)$ converges uniformly on B to $f + g$.   □

(ii)   This is **False**.

For each $n \in \mathbb{N}^+$, define $f_n, g_n \colon \mathbb{R} \to \mathbb{R}$ by $f_n(x) = g_n(x) = x + \frac{1}{n}$ and define $f, g \colon \mathbb{R} \to \mathbb{R}$ by $f(x) = g(x) = x$. We first show that (f_n) converges uniformly on $\mathbb{R}$ to f. To this end, let $\epsilon > 0$. By the Archimedean Property of $\mathbb{R}$ (Theorem 6.16), we can choose $K \in \mathbb{N}^+$ with $K > \frac{1}{\epsilon}$, or equivalently, $\frac{1}{K} < \epsilon$.

Suppose that $x \in \mathbb{R}$ and $n > K$. Then we have

$$|f_n(x) - f(x)| = \left| x + \frac{1}{n} - x \right| = \frac{1}{n} < \frac{1}{K} < \epsilon.$$

So, (f_n) converges uniformly on $\mathbb{R}$ to f. Since $(g_n) = (f_n)$ and $g = f$, we also have that (g_n) converges uniformly on $\mathbb{R}$ to g.

Now, for each $n \in \mathbb{N}^+$ and $x \in \mathbb{R}$, we have $(f_n g_n)(x) = \left(x + \frac{1}{n} \right)\left(x + \frac{1}{n} \right) = x^2 + \frac{2}{n}x + \frac{1}{n^2}$. $(f_n g_n)$ converges pointwise to the function $fg \colon \mathbb{R} \to \mathbb{R}$ defined by $(fg)(x) = x^2$. We will now show that this convergence is **not** uniform.

To this end, let $\epsilon = 2$ and for each $n \in \mathbb{N}^+$, let $x_n = n$. Then we have

$$|(f_n g_n)(n) - (fg)(n)| = \left| n^2 + \frac{2}{n}n + \frac{1}{n^2} - n^2 \right| = \left| 2 + \frac{1}{n^2} \right| = 2 + \frac{1}{n^2} \ge 2.$$

By Note 6 preceding Example 16.3, $(f_n g_n)$ does **not** converge uniformly on $\mathbb{R}$ to fg.   □

7. Find the Maclaurin series $T_f(x)$ for $f(x) = \sin x$ and show that $f(x) = T_f(x)$ for all $x \in \mathbb{R}$.

Solution: Let's compute the Taylor polynomials for f at $x = 0$. We have

$$f(x) = \sin x \qquad\qquad f(0) = 0$$
$$f'(x) = \cos x \qquad\qquad f'(0) = 1$$
$$f''(x) = -\sin x \qquad\qquad f''(0) = 0$$
$$f'''(x) = -\cos x \qquad\qquad f'''(0) = -1$$

The pattern of derivatives repeats beginning with $f^{(4)}(x)$, and so, for each $k \in \mathbb{N}$, we have $f^{(2k)}(0) = 0$ and $f^{(2k+1)}(0) = (-1)^k$.

So, we have the following:

$$p_0(x) = f(0) = 0$$
$$p_1(x) = f(0) + f'(0)x = x$$
$$p_2(x) = f(0) + f'(0)x + \frac{f''(0)}{2!}x^2 = x$$
$$p_3(x) = f(0) + f'(0)x + \frac{f''(0)}{2!}x^2 + \frac{f'''(0)}{3!}x^3 = x - \frac{1}{3!}x^3$$

In general, for each $k \in \mathbb{N}$, we have

$$p_{2k+1}(x) = \sum_{n=0}^{k} \frac{(-1)^n}{(2n+1)!}x^{2n+1} = x - \frac{x^3}{3!} + \frac{x^5}{5!} - \cdots + \frac{(-1)^n}{(2n+1)!}x^{2n+1}$$

$$p_{2k+2}(x) = \sum_{n=0}^{k} \frac{(-1)^n}{(2n+1)!}x^{2n+1} = x - \frac{x^3}{3!} + \frac{x^5}{5!} - \cdots + \frac{(-1)^n}{(2n+1)!}x^{2n+1}.$$

It follows that

$$T_f(x) = \sum_{n=0}^{\infty} \frac{(-1)^n}{(2n+1)!}x^{2n+1} = x - \frac{x^3}{3!} + \frac{x^5}{5!} - \cdots + \frac{(-1)^n}{(2n+1)!}x^{2n+1} + \cdots.$$

Since $|\sin(x)| \leq 1$ and $|\cos x| \leq 1$ for all $x \in \mathbb{R}$, the remainder term for $p_k(x)$ satisfies

$$0 \leq |R_k(x)| \leq \frac{|x|^{k+1}}{(k+1)!}.$$

By the Ratio Test for Sequences (Problem 18 from Problem Set 14), $\lim\limits_{k \to \infty} \frac{|x|^{k+1}}{(k+1)!} = 0$ (Check this!). By the Squeeze Theorem (Problem 9 in Problem Set 14), $\lim\limits_{k \to \infty} R_k(x) = 0$. Therefore, $f(x) = T_f(x)$ for all $x \in \mathbb{R}$. $\qquad\square$

8. Let $B \subseteq A \subseteq \mathbb{R}$ and for each $n \in \mathbb{N}$, let $f_n : A \to \mathbb{R}$ be a function that is bounded on B. We say that (f_n) is **uniformly bounded on B** if there is $M \in \mathbb{R}^+$ such that $|f_n(x)| \leq M$ for all $n \in \mathbb{N}$ and all $x \in B$. Prove each of the following:

   (i)   If (f_n) converges uniformly on B to $f : B \to \mathbb{R}$, then f is bounded on B.

   (ii)   If (f_n) converges uniformly on B, then (f_n) is uniformly bounded on B.

   (iii) If (f_n) is uniformly bounded on B and (f_n) converges pointwise to $f : B \to \mathbb{R}$, then f is bounded on B.

Proofs:

(i)   For each $n \in \mathbb{N}$, there is $M_n \in \mathbb{R}^+$ such that $|f_n(x)| \leq M_n$ for all $x \in B$ (because each f_n is bounded on B). Since (f_n) converges uniformly on B to f, there is $K \in \mathbb{N}$ such that for all $x \in B$, if $n > K$, then $|f_n(x) - f(x)| < 1$. In particular, for all $x \in B$,

$$|f_{K+1}(x) - f(x)| < 1 \qquad \text{and} \qquad |f_{K+1}(x)| \leq M_{K+1}.$$

So, if $x \in B$, we have

$$|f(x)| = |f(x) - f_{K+1}(x) + f_{K+1}(x)|$$
$$\leq |f(x) - f_{K+1}(x)| + |f_{K+1}(x)| \leq 1 + M_{K+1}.$$

Therefore, f is bounded on B by $1 + M_{K+1}$. □

(ii)   For each $n \in \mathbb{N}$, there is $M_n \in \mathbb{R}^+$ such that $|f_n(x)| \leq M_n$ for all $x \in B$ (because each f_n is bounded on B). Since (f_n) converges uniformly on B to f, there is $K \in \mathbb{N}$ such that for all $x \in B$, if $n > K$, then $|f_n(x) - f(x)| < 1$. So, for all $x \in B$, if $n > K$, we have

$$|f_n(x)| = |f_n(x) - f(x) + f(x) - f_{K+1}(x) + f_{K+1}(x)|$$
$$\leq |f_n(x) - f(x)| + |f(x) - f_{K+1}(x)| + |f_{K+1}(x)| \leq 1 + 1 + M_{K+1} = 2 + M_{K+1}.$$

Let $M = \max\{M_0, M_1, M_2, \ldots, M_K, 2 + M_{K+1}\}$. Then for all $x \in B$ and $n \in \mathbb{N}$, $|f_n(x)| \leq M$. Therefore, (f_n) is uniformly bounded on B. □

(iii) Since (f_n) is uniformly bounded on B, there is $M \in \mathbb{R}^+$ such that $|f_n(x)| \leq M$ for all $n \in \mathbb{N}$ and all $x \in B$. We will show that $|f(x)| \leq M$ for all $x \in B$.

   Let $x \in B$ and let $\epsilon > 0$. Since (f_n) converges to f, there is $K \in \mathbb{N}$ such that $n > K$ implies $|f_n(x) - f(x)| < \epsilon$. In particular, we have

$$|f_{K+1}(x) - f(x)| < \epsilon \qquad \text{and} \qquad |f_{K+1}(x)| \leq M.$$

   So,

$$|f(x)| = |f(x) - f_{K+1}(x) + f_{K+1}(x)| \leq |f(x) - f_{K+1}(x)| + |f_{K+1}(x)| < \epsilon + M.$$

   Since $\epsilon > 0$ was arbitrary, $|f(x)| \leq M$. Therefore, f is bounded on B by M. □

9. Find the Taylor series $T_f(x)$ for $f(x) = \ln x$ at $x = 1$ and show that $f(x) = T_f(x)$ for all x inside the radius of convergence of $T_f(x)$.

Solution: By part 3 of Example 16.25, we have

$$\frac{1}{x} = T_h(x) = \sum_{n=0}^{\infty} (-1)^n (x-1)^n = 1 - (x-1) + (x-1)^2 - \cdots + (-1)^k (x-1)^k + \cdots,$$

where $T_h(x)$ is the Taylor expansion of $h(x) = \frac{1}{x}$ at $x = 1$ with radius of convergence 1 and interval of convergence $(0, 2)$.

By Corollary 16.18, we have

$$\ln x = \int_1^x \frac{1}{t}\, dt = \int_1^x \sum_{n=0}^{\infty} (-1)^n (t-1)^n\, dt = \sum_{n=0}^{\infty} \int_1^x (-1)^n (t-1)^n\, dt$$

$$= \sum_{n=0}^{\infty} \frac{(-1)^n (x-1)^{n+1}}{n+1} = \sum_{n=1}^{\infty} \frac{(-1)^{n-1}(x-1)^n}{n}$$

$$= (x-1) - \frac{(x-1)^2}{2} + \frac{(x-1)^3}{3} - \cdots + \frac{(-1)^{k-1}(x-1)^k}{k} + \cdots,$$

By part 4 of Example 16.23, $\ln x = T_f(x)$ for all $x \in (0, 2)$.

Now, if we let $s_n = \frac{(-1)^{n-1}(x-1)^n}{n}$, then we have

$$\lim_{n\to\infty} \left| \frac{s_{n+1}}{s_n} \right| = \lim_{n\to\infty} \left| \frac{\frac{(-1)^n (x-1)^{n+1}}{n+1}}{\frac{(-1)^{n-1}(x-1)^n}{n}} \right| = \lim_{n\to\infty} \left| \frac{(-1)^n (x-1)^{n+1}}{n+1} \cdot \frac{n}{(-1)^{n-1}(x-1)^n} \right|$$

$$= \lim_{n\to\infty} \left(\left| \frac{(-1)^n}{(-1)^{n-1}} \right| \left| \frac{n}{n+1} \right| \left| \frac{(x-1)^{n+1}}{(x-1)^n} \right| \right) = |x-1| \lim_{n\to\infty} \frac{n}{n+1} = |x-1|.$$

By the Ratio Test (Theorem 15.21), if $|x-1| < 1$, the series is absolutely convergent and if $|x-1| > 1,$, the series is divergent. So, the radius of convergence of this power series is 1. Therefore, $f(x) = T_f(x)$ for all x inside the radius of convergence of $T_f(x)$.

Note: When $x = 0$, $T_f(0)$ is a multiple of the divergent harmonic series, and therefore, divergent. When $x = 2$, $T_f(2)$ is the convergent alternating harmonic series, which by Problem 15 in Problem Set 15, is equal to $\ln 2$. Therefore, the interval of convergence of $T_f(x)$ is $(0, 2]$ and $f(x) = T_f(x)$ for all x in the interval of convergence of $T_f(x)$.

10. Let (f_n) be a sequence of continuous functions on $A \subseteq \mathbb{R}$, let $B \subseteq A$, and suppose that (f_n) converges uniformly on B to f. For each $n \in \mathbb{N}$, let $x_n \in B$ and let $x \in B$ with $x_n \to x$. Prove that $\lim_{k\to\infty} \left(\lim_{n\to\infty} f_n(x_k) \right) = \lim_{n\to\infty} \left(\lim_{k\to\infty} f_n(x_k) \right)$. Is the result still true if we replace the word "uniformly" by the word "pointwise?"

Proof: Since (f_n) converges to f on B, for each $k \in \mathbb{N}$, $\lim\limits_{n\to\infty} f_n(x_k) = f(x_k)$. Since the convergence is uniform, by Theorem 16.7, f is continuous on B. Therefore, by Theorem 9.1 (1→3),

$$\lim_{k\to\infty} \left(\lim_{n\to\infty} f_n(x_k) \right) = \lim_{k\to\infty} f(x_k) = f(x).$$

Since each f_n is continuous on B, for each $n \in \mathbb{N}$, again by Theorem 9.1 (1→3), $\lim\limits_{k\to\infty} f_n(x_k) = f_n(x)$. Since (f_n) converges to f on B,

$$\lim_{n\to\infty} \left(\lim_{k\to\infty} f_n(x_k) \right) = \lim_{n\to\infty} f_n(x) = f(x).$$

Since $\lim\limits_{k\to\infty} \left(\lim\limits_{n\to\infty} f_n(x_k) \right)$ and $\lim\limits_{n\to\infty} \left(\lim\limits_{k\to\infty} f_n(x_k) \right)$ are both equal to $f(x)$, they are equal to each other.

For each $n \in \mathbb{N}$, let $f_n \colon \mathbb{R} \to \mathbb{R}$ be defined by $f_n(x) = x^n$. Let $f(x) = \begin{cases} 0 & \text{if } -1 < x < 1. \\ 1 & \text{if } \quad x = 1. \end{cases}$ Then (f_n) converges to f pointwise on $(-1, 1]$. Consider the sequence $(x_k) = 1 - \frac{1}{k}$. We have the following:

$$\lim_{k\to\infty} \left(\lim_{n\to\infty} f_n(x_k) \right) = \lim_{k\to\infty} \left(\lim_{n\to\infty} f_n\left(1 - \frac{1}{k}\right) \right) = \lim_{k\to\infty} \left(\lim_{n\to\infty} \left(1 - \frac{1}{k}\right)^n \right) = \lim_{k\to\infty} 0 = 0$$

$$\lim_{n\to\infty} \left(\lim_{k\to\infty} f_n(x_k) \right) = \lim_{n\to\infty} \left(\lim_{k\to\infty} f_n\left(1 - \frac{1}{k}\right) \right) = \lim_{n\to\infty} \left(\lim_{k\to\infty} \left(1 - \frac{1}{k}\right)^n \right) = \lim_{n\to\infty} 1^n = 1$$

So, $\lim\limits_{k\to\infty} \left(\lim\limits_{n\to\infty} f_n(x_k) \right) \neq \lim\limits_{n\to\infty} \left(\lim\limits_{k\to\infty} f_n(x_k) \right)$. Therefore, if we replace the word "uniformly" by the word "pointwise," then the statement is **false**. $\square$

11. Define $f_0 \colon \mathbb{R} \to \mathbb{R}$ by $f_0(x) = 1$ and for each $n \in \mathbb{N}$, define $f_{n+1} \colon \mathbb{R} \to \mathbb{R}$ recursively by $f_{n+1}(x) = 1 + \int_0^x f_n(t)\, dt$. Prove that the sequence (f_n) converges to a function $f \colon \mathbb{R} \to \mathbb{R}$ such that the convergence is uniform on any closed interval and $f(x) = e^x$ for all $x \in \mathbb{R}$.

Proof: We prove by induction on $n \in \mathbb{N}$ that $f_n = p_n$, where p_n is the nth Taylor polynomial of $f(x) = e^x$.

Base case ($k = 0$): In part 2 of Example 16.23, we showed that $p_0(x) = 1 = f_0(x)$.

Inductive step: Let $k \in \mathbb{N}$ and assume that $f_k(x) = p_k(x)$. Again, by part 2 of Example 16.23, we have

$$f_{n+1}(x) = 1 + \int_0^x f_n(t)\, dt = 1 + \int_0^x p_n(t)\, dt = 1 + \int_0^x \sum_{n=0}^{k} \frac{t^n}{n!}\, dt = 1 + \sum_{n=0}^{k} \frac{x^{n+1}}{(n+1)!}$$

$$= 1 + \sum_{n=1}^{k+1} \frac{x^n}{n!} = \sum_{n=0}^{k+1} \frac{x^n}{n!} = p_{k+1}(x).$$

Consider the Taylor series

$$T_f(x) = \sum_{n=0}^{\infty} \frac{x^n}{n!}$$

By part 2 of Example 16.15, the radius of convergence of this series is ∞. By Theorem 16.16 and Note 1 below the statement of Theorem 16.16, $(f_n) = (p_n)$ converges uniformly on any closed interval to $T_f(x)$.

By part 2 of Example 16.25, $T_f(x) = f(x) = e^x$. □

LEVEL 5

12. Define $f: \mathbb{R} \to \mathbb{R}$ by $f(x) = \begin{cases} e^{-\frac{1}{x^2}} & \text{if } x \neq 0. \\ 0 & \text{if } x = 0. \end{cases}$ Prove that f is infinitely differentiable at every $x \in \mathbb{R}$, but $f \neq T_f$, where $T_f(x)$ is the Maclaurin expansion for f.

Proof: We will first prove two lemmas:

Lemma 1: For each $n \in \mathbb{N}$, there is a polynomial g such that $f^{(n)}(x) = e^{-\frac{1}{x^2}} g\left(\frac{1}{x}\right)$ for all $x \in \mathbb{R}^*$.

Proof of Lemma 1: We prove this by induction.

Base case ($k = 0$): If $x \in \mathbb{R}^*$, then $f(x) = e^{-\frac{1}{x^2}} \cdot 1$ and 1 is a polynomial.

Inductive step: Let $k \in \mathbb{N}$ and assume that $f^{(k)}(x) = e^{-\frac{1}{x^2}} g\left(\frac{1}{x}\right)$. Then we have

$$f^{(k+1)}(x) = e^{-\frac{1}{x^2}} g'\left(\frac{1}{x}\right)\left(-\frac{1}{x^2}\right) + g\left(\frac{1}{x}\right) e^{-\frac{1}{x^2}}\left(\frac{2}{x^3}\right)$$

$$= e^{-\frac{1}{x^2}}\left[-\frac{1}{x^2} g'\left(\frac{1}{x}\right) + \frac{2}{x^3} g\left(\frac{1}{x}\right)\right] = e^{-\frac{1}{x^2}}\left[-\left(\frac{1}{x}\right)^2 g'\left(\frac{1}{x}\right) + 2\left(\frac{1}{x}\right)^3 g\left(\frac{1}{x}\right)\right] = e^{-\frac{1}{x^2}} h\left(\frac{1}{x}\right),$$

where $h(x) = -x^2 g'(x) + 2x^3 g(x)$. Since the derivative of a polynomial is a polynomial, the product of two polynomials is a polynomial, and the sum of two polynomials is a polynomial, h is a polynomial.

By the Principle of Mathematical Induction, for each $n \in \mathbb{N}$, there is a polynomial g such that $f^{(n)}(x) = e^{-\frac{1}{x^2}} g\left(\frac{1}{x}\right)$ for all $x \in \mathbb{R}^*$. □

Lemma 2: For each $n \in \mathbb{N}$, $f^{(n)}(0) = 0$.

Proof of Lemma 2: We prove this by induction as well.

Base case ($k = 0$): $f^{(0)}(0) = f(0) = 0$ by definition.

Inductive step: Let $n \in \mathbb{N}$ and assume that $f^{(n)}(0) = 0$. By Lemma 1, there is a polynomial g such that $f^{(n)}(x) = e^{-\frac{1}{x^2}} g\left(\frac{1}{x}\right)$ for all $x \in \mathbb{R}^*$. Letting $t = \frac{1}{x}$, we have

$$f^{(n+1)}(0) = \lim_{x \to 0} \frac{f^{(n)}(x) - f^{(n)}(0)}{x - 0} = \lim_{x \to 0} \frac{e^{-\frac{1}{x^2}} g\left(\frac{1}{x}\right)}{x} = \lim_{t \to \infty} \frac{tg(t)}{e^{t^2}} = 0.$$

By the Principle of Mathematical Induction, for each $n \in \mathbb{N}$, $f^{(n)}(0) = 0$. □

Proof of Main Result: Since $\frac{1}{x}, -\frac{1}{x^2}, e^x$, and all polynomials are differentiable at every $x \in \mathbb{R}^*$, $e^{-\frac{1}{x^2}} g\left(\frac{1}{x}\right)$ is differentiable at every $x \in \mathbb{R}^*$ for any polynomial g. By Lemma 1, f is infinitely differentiable at every $x \in \mathbb{R}^*$. By Lemma 2, f is infinitely differentiable at $x = 0$. So, f is infinitely differentiable at every $x \in \mathbb{R}$. By Lemma 2, $T_f(x) = 0$ for all $x \in \mathbb{R}$. Since $e^{-\frac{1}{x^2}} > 0$ for all $x \neq 0$, it follows that $f \neq T_f$. □

13. A sequence of functions (f_n) is said to be **increasing** on A if for all $x \in A$ and $n \in \mathbb{N}$, $f_n(x) \leq f_{n+1}(x)$. Similarly, a sequence of functions (f_n) is said to be **decreasing** on A if for all $x \in A$ and $n \in \mathbb{N}$, $f_n(x) \geq f_{n+1}(x)$. Finally, a sequence of functions (f_n) is said to be **monotone** on A if the sequence is increasing or decreasing on A.

Let (f_n) be a monotone sequence of continuous functions on $A \subseteq \mathbb{R}$, let $[a, b] \subseteq A$, and suppose that (f_n) converges on $[a, b]$ to a continuous function $f : [a, b] \to \mathbb{R}$. Prove that (f_n) converges uniformly on $[a, b]$ to f. This result is known as **Dini's Theorem**.

Proof: First assume that (f_n) is an increasing sequence of continuous functions on A. Let $\epsilon > 0$ and for each $n \in \mathbb{N}$, let $g_n = f - f_n$. Then (g_n) is a decreasing sequence of continuous functions on A. For each $n \in \mathbb{N}$, let $V_n = g_n^{-1}(-\infty, \epsilon) = \{x \in [a, b] \mid g_n(x) < \epsilon\}$. By Theorem 9.3, for each $n \in \mathbb{N}$, there is an open set $U_n \subseteq \mathbb{R}$ such that $V_n = U_n \cap [a, b]$. Since (g_n) is decreasing, for each $n \in \mathbb{N}$, we have $V_n \subseteq V_{n+1}$. Since $g_n \to 0$, the collection $\mathcal{C} = \{U_n \mid n \in \mathbb{N}\}$ is an open cover of $[a, b]$. By the Heine-Borel Theorem (Theorem 7.28), $[a, b]$ is compact. Therefore, there is a finite subcollection $\mathcal{D} \subseteq \mathcal{C}$ such that $\mathcal{D}$ is still an open cover of $[a, b]$. If we let $K = \max\{n \in \mathbb{N} \mid U_n \in \mathcal{C}\}$, then $V_K = U_K \cap [a, b] = [a, b]$. So, if $x \in [a, b]$ and $n > K$, then $g_n(x) < \epsilon$, or equivalently, $|f_n - f| = f - f_n < \epsilon$. Since $\epsilon > 0$ was arbitrary, (f_n) converges uniformly on $[a, b]$ to f.

Next, assume that (f_n) is a decreasing sequence of continuous functions on A that converges on $[a, b]$ to a continuous function f. Then $(-f_n)$ is an increasing sequence of continuous function on A that converges on $[a, b]$ to the continuous function $-f$. By the first paragraph above, $(-f_n)$ converges uniformly on $[a, b]$ to $-f$. Let $\epsilon > 0$. Since $(-f_n)$ converges uniformly on $[a, b]$ to $-f$, there is $K \in \mathbb{N}$ such that if $x \in [a, b]$ and $n > K$, then $|f_n - f| = |-(f_n - f)| = |-f_n - (-f)| < \epsilon$. Therefore, f_n converges uniformly on $[a, b]$ to f. □

About the Author

Dr. Steve Warner, a New York native, earned his Ph.D. at Rutgers University in Pure Mathematics in May 2001. While a graduate student, Dr. Warner won the TA Teaching Excellence Award.

After Rutgers, Dr. Warner joined the Penn State Mathematics Department as an Assistant Professor and in September 2002, he returned to New York to accept an Assistant Professor position at Hofstra University. By September 2007, Dr. Warner had received tenure and was promoted to Associate Professor. He has taught undergraduate and graduate courses in Precalculus, Calculus, Linear Algebra, Differential Equations, Mathematical Logic, Set Theory, and Abstract Algebra.

From 2003 – 2008, Dr. Warner participated in a five-year NSF grant, "The MSTP Project," to study and improve mathematics and science curriculum in poorly performing junior high schools. He also published several articles in scholarly journals, specifically on Mathematical Logic.

Dr. Warner has nearly two decades of experience in general math tutoring and tutoring for standardized tests such as the SAT, ACT, GRE, GMAT, and AP Calculus exams. He has tutored students both individually and in group settings.

In February 2010 Dr. Warner released his first SAT prep book "The 32 Most Effective SAT Math Strategies," and in 2012 founded Get 800 Test Prep. Since then Dr. Warner has written books for the SAT, ACT, SAT Math Subject Tests, AP Calculus exams, and GRE. In 2018 Dr. Warner released his first pure math book called "Pure Mathematics for Beginners." Since then he has released several more books, each one addressing a specific subject in pure mathematics.

Dr. Steve Warner can be reached at

steve@SATPrepGet800.com

BOOKS BY DR. STEVE WARNER

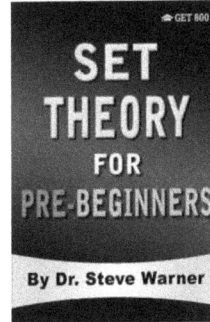

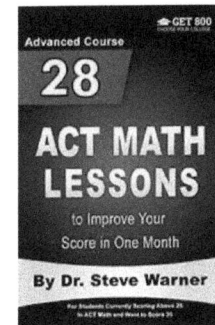

PURE MATHEMATICS FOR PRE-BEGINNERS
Logic
Set Theory
Abstract Algebra
Number Theory
Real Analysis
Topology
Complex Analysis
Linear Algebra
By Dr. Steve Warner

PURE MATHEMATICS FOR BEGINNERS
Logic
Set Theory
Abstract Algebra
Number Theory
Real Analysis
Topology
Complex Analysis
Linear Algebra
By Dr. Steve Warner

SET THEORY FOR PRE-BEGINNERS
By Dr. Steve Warner

SET THEORY FOR BEGINNERS
By Dr. Steve Warner

TOPOLOGY FOR BEGINNERS
By Dr. Steve Warner

ABSTRACT ALGEBRA FOR BEGINNERS
By Dr. Steve Warner

REAL ANALYSIS FOR BEGINNERS
By Dr. Steve Warner

1000 NEW SAT MATH PROBLEMS arranged by Topic and Difficulty Level
By Dr. Steve Warner
1000 Problems with Full Explanations for the New SAT

Second Edition
320 AP CALCULUS AB PROBLEMS arranged by Topic and Difficulty Level
By Dr. Steve Warner

Second Edition
320 AP CALCULUS BC PROBLEMS arranged by Topic and Difficulty Level
By Dr. Steve Warner

LEVEL 1
320 SAT MATH SUBJECT TEST PROBLEMS arranged by Topic and Difficulty Level
By Dr. Steve Warner

LEVEL 2
320 SAT MATH SUBJECT TEST PROBLEMS arranged by Topic and Difficulty Level
By Dr. Steve Warner

Second Edition
320 ACT MATH PROBLEMS arranged by Topic and Difficulty Level
By Dr. Steve Warner

Beginner Course
28 ACT MATH LESSONS to Improve Your Score in One Month
By Dr. Steve Warner

Intermediate Course
28 ACT MATH LESSONS to Improve Your Score in One Month
By Dr. Steve Warner

Advanced Course
28 ACT MATH LESSONS to Improve Your Score in One Month
By Dr. Steve Warner

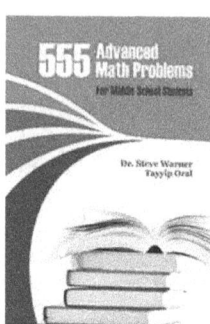

www.ingramcontent.com/pod-product-compliance
Lightning Source LLC
Chambersburg PA
CBHW081815200326
41597CB00023B/4264